AF389450

Challenges and Opportunities for the Renewable Energy Economy

Challenges and Opportunities for the Renewable Energy Economy

Special Issue Editor

Juan Carlos Reboredo

MDPI • Basel • Beijing • Wuhan • Barcelona • Belgrade

Special Issue Editor
Juan Carlos Reboredo
University of Santiago de
Compostela
Spain

Editorial Office
MDPI
St. Alban-Anlage 66
4052 Basel, Switzerland

This is a reprint of articles from the Special Issue published online in the open access journal *Energies* (ISSN 1996-1073) from 2019 to 2020 (available at: https://www.mdpi.com/journal/energies/special_issues/challenges_and_opportunities_for_the_renewable_energy_economy).

For citation purposes, cite each article independently as indicated on the article page online and as indicated below:

LastName, A.A.; LastName, B.B.; LastName, C.C. Article Title. *Journal Name* **Year**, *Article Number*, Page Range.

ISBN 978-3-03936-473-2 (Hbk)
ISBN 978-3-03936-474-9 (PDF)

Contents

About the Special Issue Editor . . . . **vii**

Preface to "Challenges and Opportunities for the Renewable Energy Economy" . . . . **ix**

Chao Bi, Jingjing Zeng, Wanli Zhang and Yonglin Wen
Modelling the Coevolution of the Fuel Ethanol Industry, Technology System, and Market
System in China: A History-Friendly Model
Reprinted from: *Energies* **2020**, *13*, 1034, doi:10.3390/en13051034 . . . . **1**

Mohd Amin Abd Majid, Hamdan Haji Ya, Othman Mamat and Shuhaimi Mahadzir
Techno Economic Evaluation of Cold Energy from Malaysian Liquefied Natural Gas
Regasification Terminals
Reprinted from: *Energies* **2019**, *12*, 4475, doi:10.3390/en12234475 . . . . **27**

Juan C. Reboredo, Andrea Ugolini and Yinfei Chen
Interdependence Between Renewable-Energy and Low-Carbon Stock Prices
Reprinted from: *Energies* **2019**, *12*, 4461, doi:10.3390/en12234461 . . . . **41**

Hussein M. K. Al-Masri, Ayman Al-Quraan, Ahmad AbuElrub and Mehrdad Ehsani
Optimal Coordination of Wind Power and Pumped Hydro Energy Storage
Reprinted from: *Energies* **2019**, *12*, 4387, doi:10.3390/en12224387 . . . . **55**

Paulino Martinez-Fernandez, Fernando deLlano-Paz, Anxo Calvo-Silvosa and Isabel Soares
Assessing Renewable Energy Sources for Electricity (RES-E) Potential Using a
CAPM-Analogous Multi-Stage Model
Reprinted from: *Energies* **2019**, *12*, 3599, doi:10.3390/en12193599 . . . . **70**

**Tengku Adeline Adura Tengku Hamzah, Zainorfarah Zainuddin, Mariney Mohd Yusoff,
Saripah Osman, Alias Abdullah, Khairos Md Saini and Arno Sisun**
The Conundrum of Carbon Trading Projects towards Sustainable Development: A Review from
the Palm Oil Industry in Malaysia
Reprinted from: *Energies* **2019**, *12*, 3530, doi:10.3390/en12183530 . . . . **90**

Anna Manowska and Andrzej Nowrot
The Importance of Heat Emission Caused by Global Energy Production in Terms of
Climate Impact
Reprinted from: *Energies* **2019**, *12*, 3069, doi:10.3390/en12163069 . . . . **105**

**Yaovi Ouézou Azouma, Lynn Drigalski, Zdeněk Jegla, Marcus Reppich, Vojtěch Turek and
Maximilian Weiß**
Indirect Convective Solar Drying Process of Pineapples as Part of Circular Economy Strategy
Reprinted from: *Energies* **2019**, *12*, 2841, doi:10.3390/en12152841 . . . . **117**

Hugo Algarvio, António Couto, Fernando Lopes and Ana Estanqueiro
Changing the Day-Ahead Gate Closure to Wind Power Integration: A Simulation-Based Study
Reprinted from: *Energies* **2019**, *12*, 2765, doi:10.3390/en12142765 . . . . **135**

About the Special Issue Editor

Juan Carlos Reboredo is Professor of Economics at the Universidade de Santiago de Compostela, Spain. His research focuses on the analysis of risk in capital markets and the dependence relationship between financial and energy markets. Within these areas, he analyzes the influence of oil prices on the valuation of financial assets and exchange rates, systemic risk in financial markets and systemic oil market effects of renewable energy markets. His most recent research analyzes the impact of climate risk on financial and energy markets, financial stability and investment decisions. His research papers have been published in journals such as *Energy Economics, Energy Policy, Journal of Banking and Finance, Journal of International Money and Finance, Renewable and Sustainable Energy Reviews* and *Emerging Markets Review*, among others.

Preface to "Challenges and Opportunities for the Renewable Energy Economy"

Since the 2015 Paris Agreement to keep the average global temperature rise below 2 °C, investors, consumers and regulators have no longer been able to turn their backs on the need to green the economy in the face of the swelling tide of climate-related regulations and technological disruption. While the transition to a low-carbon economy might potentially cause severe disruption and losses for companies with business models that rely directly or indirectly on fossil fuels, it also brings new investment opportunities for low-carbon and renewable-energy companies.

Renewable energy deployment as an alternative to traditional energy sources is the cornerstone of emission-reduction energy policies that aim to facilitate the transition to a low-carbon economy. However, the transition to renewable energies poses risks and opportunities for companies with business models that rely directly or indirectly on renewables, and this, in turn, will be reflected in the revaluation of assets and in the reallocation of private and public financial investments from carbon-intensive to low-carbon energies.

In its nine independent chapters, this book addresses current challenges and opportunities for renewable energies from the technological, economic and financial perspectives.

Chapter One examines how entry regulations, production subsidies, R&D subsidies and the ethanol mandates impact the growth of the ethanol fuel industry in China, with specific attention paid to the effectiveness of different policies in boosting ethanol production. Chapter Two evaluates the technical and economic viability of capturing cold energy released during regasification processes in Malaysian liquefied natural gas regasification terminals, documenting that a substantial amount of cold energy could be generated with a high internal rate of return over the long term. Chapter Three explores market interdependence and price spillovers between renewable-energy and low-carbon assets so as to determine the potential diversification benefits of renewable energy investment in the European and USA stock markets. Empirical evidence shows that renewable-energy and low-carbon stock price dependence differs across markets, and this, in turn, has implications for the design of carbon-resilient portfolios and risk management strategies, as well as for the implementation of public funding policies to support the transition to a low-carbon economy. Chapter Four evaluates the advantages of combining wind power with pumped hydro-energy storage and reports that, from an economic, environmental and technical perspective, combining these two sources of energy generation is more efficient than the conventional approach to power generation. Chapter Five, considering the aim of achieving progressive decarbonization, assesses the role played by renewable energies in the power generation portfolio, considering technological and environmental restrictions, while confirming the relevance of small and large hydro and offshore wind projects as preferential technologies in efficient and diversified portfolios. Chapter Six analyzes how the introduction of sustainability-focused carbon trading projects in the Malaysian palm oil industry report beneficial effects on sustainability, investment and economic growth. Chapter Seven addresses how heat emissions from energy production contribute to the greenhouse effect by considering global heat production compared with total solar energy. Chapter Eight presents an industrial application of convective solar drying of pineapples as a circular economy device that verifies that fossil fuel consumption can be considerably reduced with the application of convective solar pre-drying processes. Finally, Chapter Nine, a simulation study that analyzes wind power forecasts and their impact on market-clearing prices, shows that enhanced forecast precision has favorable effects on market participants and on the energy system.

This book intends to be a reference work for those who approach the economic and technological features of renewable energy deployment from a research, practical or policy-making point of view. My sincere gratitude is extended to all the contributing authors for their efforts in making this book possible, with special thanks to MDPI for its continuous support and encouragement.

Juan Carlos Reboredo
Special Issue Editor

Article

Modelling the Coevolution of the Fuel Ethanol Industry, Technology System, and Market System in China: A History-Friendly Model

Chao Bi [1], Jingjing Zeng [2,*], Wanli Zhang [3] and Yonglin Wen [2]

[1] International Business School, Shaanxi Normal University, Xi'an 710119, China; bichao@snnu.edu.cn
[2] School of Public Administration, Zhongnan University of Economics and Law, Wuhan 430073, China; ylwen@zuel.edu.cn
[3] School of Economics and Finance, Xi'an Jiaotong University, Xi'an 710061, China; zhangwanli623@stu.xjtu.edu.cn
* Correspondence: jjzeng@zuel.edu.cn

Received: 14 October 2019; Accepted: 24 February 2020; Published: 26 February 2020

Abstract: The interaction among the fuel ethanol industry, the technology system, and the market system has a substantial effect on the growth of the fuel ethanol industry which plays a key role in the formation of a sustainable energy system in China. However, we know little about the relationships among them and it is difficult to explore the nexus using econometric method due to the lack of statistics on China's fuel ethanol industry. This paper develops a history-friendly coevolutionary model to describe the relationships among the fuel ethanol industry, the technology system, and the market system in China. Based on the coevolutionary model, we further assess the impacts of entry regulations, production subsidies, R&D subsidies, and ethanol mandates on the growth of the fuel ethanol industry in China using a simulation method. The results of historical replication runs show that the model can appropriately reflect the multidirectional causalities between the fuel ethanol industry, the technology system, and the market system. We also found that entry regulation is conducive to weakening the negative economic impacts induced by the growth of the grain-based fuel ethanol industry without affecting the long-term total output of the industry; production subsidies to traditional technology firms are helpful for the expansion of the fuel ethanol industry, but they also impede technology transfer in the industry; only when firms inside the industry are not in the red can R&D subsidies promote technological progress and then further accelerate the growth of the fuel ethanol industry; the ethanol mandate has a significant impact on industrial expansion only when a production subsidy policy is implemented simultaneously. Our findings suggest that more attention could be paid to consider the cumulative effects caused by coevolutionary mechanisms when policymakers assess the effects of exogenous policies on the growth of the fuel ethanol industry. More attention also could be paid to the conditions under which these policies can work effectively.

Keywords: coevolution; the fuel ethanol industry; history-friendly model; entry regulation; ethanol mandate; production subsidy; R&D subsidy

---

1. Introduction

Fuel ethanol helps to reduce harmful emissions from vehicles, contributing to the fight against climate change and the pursuit of clean mobility [1]. Moreover, as a kind of renewable energy made from sustainable biomass materials, fuel ethanol is an ideal substitute for the non-renewable fossil fuels [2]. Thus, fuel ethanol is expected to play a key role in the formation of a sustainable energy system. China is the largest consumer of fossil fuels and attaches great importance of the development of the fuel ethanol industry. In 2018, the fuel ethanol production capacity in China reached 3.22 million tons. China has become the third-largest fuel ethanol consuming and producing country, following Brazil and the United States [3]. However, China accounted for just over 3% of global production in 2018. The development of the fuel ethanol industry in China still faces many challenges (e.g., technical uncertainty, demand uncertainty, and feedstock uncertainty) [4–6]. The Chinese government announced a new nationwide ethanol mandate that will expand the mandatory use of E10 fuel (gasoline containing 10% ethanol) from 11 trial provinces to the entire country by 2020. If China were to meet the national mandate of E10, it would require an extra 12 million tons of fuel ethanol production capacity, which is about four times that of its current production capacity. Therefore, exploring the nexus among the fuel ethanol industry, the uncertain technology system, and the uncertain market demand will be conducive to clarifying the growth mechanisms of the fuel ethanol industry and then improving public policy to accelerate the growth of the fuel ethanol industry in China.

Many scholars have already analyzed the different factors that affect the development of the fuel ethanol industry. These factors mainly include technology change [7–10], market demand [11–13], feedstocks [8,14,15], renewable energy infrastructures [16,17], energy policies [18–21], and economic, social, and environmental impacts [22–25]. These works mainly used case studies [8,10], econometric methods [9,11,12,19], and simulation methods [7,13]. Although the existing literature provides important information on the relationships between the fuel ethanol industry growth and its drivers, there are still some limitations. Firstly, most studies do not consider the adverse impacts of the fuel ethanol industry on its drivers and the interrelationships among the driving factors. The ignorance of the above interactions may lead to a misunderstanding of the growth mechanisms of the fuel ethanol industry [26]. Secondly, the fuel ethanol industry is an emerging industry in China. Therefore, the econometric methods used in the existing studies could not be applied to analyze the development of the fuel ethanol industry in China due to the lack of statistical data. Lastly, although simulation is an ideal method to quantitatively analyze the development of emerging industries, such as the fuel ethanol industry, this method is often questioned because of the subjectivity of its parameter settings [27].

In addressing these limitations, this paper employs a history-friendly evolutionary model to explore the interactions among the fuel ethanol industry and its driving factors and analyze the impacts of different policies on the evolution of the fuel ethanol industry in China. The contributions of our work are reflected in three aspects. First, we argue that there are multidirectional causalities among the fuel ethanol industry, the technology system, and the market system. Therefore, we applied a coevolutionary framework to model the above relationships. Under this framework, each party exerts selective pressures on the others, thereby affecting each other's evolution [28]. Second, we developed a history-friendly model to depict the above coevolutionary relationships related to the fuel ethanol industry in China. The parameters of the history-friendly model are set based on the historical evolutionary characteristics of the industry but not on historical statistics. Therefore, this method can be used to analyze the growth of the fuel ethanol industry, which lacks historical statistics in China. At last, we further analyzed the impacts of entry regulation, production subsidy, R&D subsidy, and ethanol mandate policy on the evolution of the fuel ethanol industry in China.

The rest of the paper is structured as follows. Section 2 reviews the driving factors that affect the growth of the fuel ethanol industry. Section 3 features a history-friendly model based on the coevolutionary framework, which describes the interactions among the fuel ethanol industry, the technology system, and the market system. In Section 4, we first run the baseline simulation to select the values of the parameters that can reflect the historical characteristics of the fuel ethanol industry and then use this model to further analyze the impacts of several typical fuel ethanol industry policies, while Section 5 contains concluding remarks and policy implications.

2. Literature Review

2.1. The Effects of Technology Change on the Evolution of the Fuel Ethanol Industry

One of the most important factors limiting the scale of the fuel ethanol industry in the United States is technological progress. Fuel ethanol could account for about 10% of total energy consumption if there is no major technological advance, but if technological progress is significant, then this proportion could rise to 20%. This proportion could further rise to 25% if the major usage of the land could be partly converted to planting energy crops [7]. The success of the fuel ethanol industry was not only due to resource advantages but also due to the advantage of technological progress [8]. In addition, the effect of technological progress on the production of fuel ethanol is adjusted by mandatory consumption policies [9].

There is no doubt that technological change has vital impacts on industry evolution. However, the origin of the fuel ethanol industry's technological progress remains controversial. One of the hypotheses is that technological advantage is derived from the accumulation of knowledge, which is one of the "learning by doing" types [29]. Another opinion is that technological progress is driven by the new energy automobile industry, especially the development of flexible fuel vehicles that began in 2003 [30,31]. In addition, agricultural output is believed to be the main source of the fuel ethanol industry, so the progress of agricultural technology has had an important impact on the development of the fuel ethanol industry [32]. Different actors (e g , ethanol producers, public and private research institutions, and government institutions) in the industrial chain could also affect technological progress [7]. Some scholars found that more attention should be paid to scientific progress, especially the impact of scientific progress on fuel ethanol technology in recent years [33].

The relationship between the progress of fuel ethanol technology and public policy is very close. However, the focus of the scientific community is different than the focus of the government. The scientific community is concerned about environmental protection and technology, while the government is concerned more about geopolitics. This difference reveals that either the scientific community needs to pay more attention to the knowledge demands of policymakers or that policymakers need to focus more on scientific and technological knowledge [10].

2.2. The Effects of Renewable Energy Policy on the Evolution of the Fuel Ethanol Industry

The development of new technologies in the fuel ethanol industry still faces various barriers [34]. In many countries such as the United States [35], governments have adopted a wide range of policies to support the development of their fuel ethanol industries, especially, to spur the new technologies. These policies (e.g., subsidies, tax cuts, mandatory consumption, and tariff protections) have had important impacts on the development of the fuel ethanol industry. If the government's subsidy policy is properly designed, it can effectively compensate for the risk difference caused by land quality difference, and further, promote the greater use of marginal land with poor quality to produce energy crops [36]. R&D support policies play an important role in the technological progress of the fuel ethanol industry, and the quantity of public R&D funds directly affects the progress of fuel ethanol technology [7].

Some scholars have investigated the impact of various renewable energy policies on industry evolution. A vertically integrated market model, including fuel ethanol, by-products, and corn, was used to analyze the social impact of the fuel ethanol industry's subsidy policies, and the results showed that subsidy policies may not lead to positive social benefits [18]. It is also found that the tariff protection and tax-cut policies of biofuels in the United States may lead to the loss of total social welfare [37].

Other scholars analyzed the distortion effect of fuel ethanol policy from the perspective of international trade. If the U.S. government were to cancel tariff protection and reduce the scope of tax cuts, ethanol imports would increase 130%, while domestic ethanol production would fall 9% [11]. The current biofuel trade protection policy in the United States not only reduces the industrial competitive advantage of corn-based ethanol but also increases dynamic learning costs, which will also reduce the international competitiveness of the future cellulosic ethanol industry [38]. No matter how high the carbon tax rate is, a single carbon tax policy without additional subsidies will not promote the evolution of the fuel ethanol industry [19].

Although many studies have noted the distortion effects of fuel ethanol policies, other scholars maintain that subsidies can help to overcome market failure [20]. Gehlhar et al. (2010) evaluated the long-term economic impact of fuel ethanol subsidies in the United States and concluded that the development of the fuel ethanol industry is conducive to lowering the dependence on oil imports, thereby contributing to the development of the overall economy. Further, the overall benefits induced by the subsidy were greater than the welfare losses [21].

However, due to the characteristics of the emerging industry, the future direction of the fuel ethanol industry policy is highly uncertain, which increases the uncertainty of the development of the fuel ethanol industry [39]. The development of the fuel ethanol industry involves issues of food security, energy security, and environmental protection. Therefore, in the process of formulating supporting policies for the fuel ethanol industry, attention should be paid to the integration of these policies [32,40].

2.3. The Effects of Market Factors on the Evolution of the Fuel Ethanol Industry

Since oil and biofuels are substitutes, and crops like corn and sugar cane are the main feedstocks of biofuels, the price of oil, corn, and sugar cane will have an important impact on the evolution of the biofuel industry. An empirical study showed that an increase in oil prices will lead to an increase of biofuel production, while the increase of corn and sugarcane price will lead to a decrease of fuel ethanol production [11]. Government subsidies for the fuel ethanol industry and biodiesel industry should be increased because uncertainty in oil prices and crop yields would affect the evolution of the fuel ethanol industry [12]. It is also found that a 30% drop in oil price would lead to a significant drop in fuel ethanol demand, and, at the same time, fuel ethanol prices would also drop significantly [13].

The cost of feedstocks was found to be the main influencing factor affecting the short-term evolution of the fuel ethanol industry. Therefore, the future R&D of the fuel ethanol industry should focus on the low cost of decomposition from lignocellulose sugar and the comprehensive utilization of lignocellulose [14]. It is also found that an insufficient understanding of feedstock cost is the main reason for the slow progress in the commercial utilization of fuel ethanol in Africa [5].

The above literature sheds important light on the relationships between the fuel ethanol industry and its driving factors. However, the ignorance of the industry's multidirectional relationships may lead to an inaccurate assessment of the relationship between the fuel ethanol industry and its drivers. Therefore, in order to understand the growth mechanism of the fuel ethanol industry in China, this paper will analyze the relationships among the fuel ethanol industry, the technology system, and the market system using a history-friendly coevolutionary model. This paper will further assess the impacts of several policies (i.e., entry regulation, production subsidy, R&D subsidy, and ethanol mandates) on the growth of the fuel ethanol industry.

3. The Model

3.1. The History-Friendly Model

As an emerging industry in China, there are limited statistics of the fuel ethanol industry. Therefore, it is difficult to analyze the mechanisms and factors affecting the fuel ethanol industry using a statistical model. In order to explore the coevolutionary relationships between the fuel ethanol industry, the technology system, and the market system, this paper employs a history-friendly evolutionary model which has been applied to many industries, including computers, DRAM chips, pharmaceuticals, semiconductors, synthetic dyes, and mobile phones and memory chips [41–46]. Scholars studying industrial dynamics generally rely quite heavily on the appreciative theory which is a body of verbal arguments representing causal explanations of observed patterns of economic phenomena [47,48]. Although the appreciative theory is an appropriate tool to characterize the main mechanisms at work, it is difficult to verify the logical consistency of the theory due to its complexity and the lack of precision of the verbal language [47]. The history-friendly model is a formal model of the appreciative theory and can overcome the above limitations of the appreciative theory [41]. It aims to analyze, in a more formal form, the influential factors and their influencing mechanisms in industry evolution, technological progress, and institutional change that have been confirmed by appreciative theory [41,49].

The construction of the history-friendly model consists of three important steps [47]. The first step is the selection of the stylized facts deserving attention from theoretical perspectives. These stylized facts mainly refer to the history and evolution of the fuel ethanol industry such as specific institutions, technological change, and market characteristics. Second, there is the choice of how to represent the selected phenomena. In this respect, the history-friendly model adopts the same basic representations used in evolutionary models. All reported models are built around four main blocks: firm behavior, technological change, market demand, and industry dynamics [50,51]. The creation, entry, exit, and technological change of the business firms affect the performance of the industry and further impact industry evolution. The third step is the manipulation and implementation of the model designed in the second step.

The history-friendly model is a type of agent-based simulation model dealing with the complexity of the economic system [52]. A typical history-friendly model has many variables and parameters. Under a wide range of parameter settings, some of the parameter settings will lead to the replication of the industry history being modelled. Importantly, "replication" here mainly refers to qualitative reproduction, not quantitative reproduction [53]. Once the model is built, there is room for wider applications such as policy analysis. The history-friendly nature is threefold. Firstly, in the process of model construction, stylized facts in industrial development are fully considered, and the relationship between variables is constructed on this basis. Secondly, the initial values of variables in the model are set based on the true values of industrial history. Thirdly, the selection of the parameters' values can qualitatively reproduce the stylized facts in industrial history.

There are two compelling reasons for using a history-friendly model in this paper. First, a history-friendly model helps us better explore the causal mechanisms in the evolution of the fuel ethanol industry. As a formal model, all the logic that drives model outcomes is explicitly represented in a history-friendly model [27]. In addition, the mechanisms built into the model are transparent which means that if the model does not work as expected, the analyst can adjust the settings of the model until the model is able to qualitatively capture the stylized facts in the appreciative theory [54]. Developing and working through a history-friendly model could bring to mind mechanisms, factors, and constraints of the industry evolution [41]. Therefore, compared with the appreciative theory, the history-friendly model is conducive to analyzing the causal mechanism of the fuel ethanol industry. Second, the model setting of the history-friendly model is transparent rather than arbitrary, so, it serves as a good starting point for further policy analysis. Comparing the influence of different systems and policy arrangements on industry evolution can provide a deep understanding of the influence mechanism of the above factors and provide a basis for further policy selection and institutional arrangement [41,55].

3.2. The Model Specification

The basic model is presented in this section. Given the complexity of the history-friendly model, it is difficult to lay out all the details of all the equations without confusing the reader and obscuring the basic logic of the model [41]. Therefore, we have tried to present only the equations which can reflect the stylized facts of the fuel ethanol industry and put other related equations in the Appendix A. Just like most of the other history-friendly models [50,51], our model is also built around four main building blocks: firm, technology progress, market demand, and industrial dynamics. In selecting the stylized facts to investigate, we considered their relevance on the selection of the variables and the relationships among these variables, which affect the model specification [47]. These stylized facts are put at the beginning of each block. In addition, given that our model involves many subjects which are further divided into different types, we use a lot of superscripts to minimize the number of variables used in the model. The variables with superscripts b and f represent the variables associated with the properties of fuel ethanol and fossil fuels, respectively. The variables with superscripts tf, nf, and rd represent the variables associated with the properties of traditional technology firms, new technology firms, and R&D firms, respectively. The variables with superscripts m and t represent the variables associated with the properties of materials and traditional materials, respectively.

3.2.1. Firm

(1) R&D investment of the firm

As a typical emerging industry, the fuel ethanol industry is still faced with the urgent need for continuous improvement of its related technologies. Therefore, one of the stylized facts is that almost all firms in the fuel ethanol industry have research and development (R&D) investments. We assume that the firm's R&D investment consists of two parts. The first part is the fixed amount of R&D investment. Whether the company is profitable or not, it will invest in R&D. The second part is that when the firm has a positive profit, a fixed proportion of profit will be invested in R&D. Thus, R&D investment can be expressed as:

$$R_{i,t} = \text{Max}\{rd + \sigma \cdot \pi_{i,t}, rd\} \tag{1}$$

where rd denotes the fixed amount of R&D investment; σ denotes the fixed proportion of profit invested in R&D, and π_{it} denotes firm's profit which is defined by Equation (A7) in the Appendix A.

(2) Entry of firms

We assume that a firm's entry decision is influenced by the industry's profit to cost ratio. Let $\varphi(x) = \Phi \cdot \exp(-\varphi \cdot x)$, where φ and Φ are positive constants, and $\Phi \in (0, 1]$ is given a distribution function p_s, $s = 1,2...,l$; then, the number of latecomers in each period can be expressed by the following equation:

$$\gamma_s = \begin{cases} 0 & \text{with probability } \psi(x) \\ s & \text{with probability } p_s \cdot (1 - \psi(x)) \end{cases} \tag{2}$$

where $\psi(x) = \varphi(\text{max}[\Gamma_t, 0])$.

In Equation (2), if $\Phi = 1$ is set; then, $\varphi(0) = 1$, which represents when the incumbent firm loses money, and no new firms enter this industry. That is to say, $\Phi = 1$ indicates that the firm is completely rational. If $\Phi < 1$ is set, even if the incumbent firms have losses, there will still be latecomers. In other words, this model can satisfy the theoretical hypothesis of rationality or incomplete rationality by making different assumptions.

This study assumes that the initial size and technical efficiency of the latecomers are equal to the average level of the whole industry.

(3) Adjustment rules of the firm

During each period, the firm can determine the optimal output, $s_{i,t}$, and the corresponding demand of feedstock, $m_{i,t}$, according to the feedstock price and product price. Due to the matching relationship between the feedstock input and fixed assets, the required asset size should be $m_{i,t}/\alpha$. If the firm's own asset scale $F_{i,t-1}$ is smaller than $m_{i,t}/\alpha$, then the firm's asset scale expands to $F_{i,t} = m_{i,t}/\alpha$, and the corresponding firm's capacity utilization ratio is $\eta_{i,t} = 1$. Otherwise, if the firm's own asset scale $F_{i,t-1}$ is larger than $m_{i,t}/\alpha$, the firm's asset scale remains unchanged, that is $F_{i,t} = F_{i,t-1}$, and the capacity utilization rate is $\eta_{i,t} = m_{i,t}/(\alpha \cdot F_{i,t})$.

(4) Exit rules of the firm

We assume that when the firm has losses for several consecutive periods, the firm will withdraw production.

3.2.2. Technology Progress

(1) Progress and Diffusion of Traditional Technology

There are four stylized facts of the technological progress in China's fuel ethanol industry. First, traditional production technology is relatively mature, so technological progress is mostly reflected in the continuous improvement of the original technology. However, there are a few major technological innovations. In other words, with an increase in the degree of technological progress, the occurrence probability of technological progress rapidly decreases. Second, R&D investment will improve the probability of technological progress. Third, the higher the original level of technology, the lower the probability of major innovation. Finally, due to technology diffusion, the technological progress of a specific firm is positively correlated with the most advanced technology level in the industry.

In this study, it is assumed that there is the highest level of technical efficiency boundary, denoted as e_0. Let $\Delta e_{i,t}$ be the change of the firm's technical level; then, the technology change will not exceed the difference between the firm's technical level and the highest level $(e_0 - e_{i,t})$. Therefore, the firm's technology change is

$$\Delta e_{i,t+1} = \theta_{i,t+1} \cdot (e_0 - e_{i,t}) \tag{3}$$

where $\theta_{i,t+1}$ is the random variable of the interval [0,1], in order to reflect the first stylized fact of technological progress, that is, the larger the degree of technological progress, the smaller the occurrence probability.

We construct variables $k = 100 \cdot \theta_{i,t+1}$ and assume Poisson distribution with parameters k and λ. Then, there is

$$\lambda = \lambda_0 \cdot R_{i,t}^{\lambda_1} \cdot (e_0 - e_{i,t})^{\lambda_2} \cdot \left(\max_i \{e_{i,t}\} / e_{i,t} \right)^{\lambda_3} \tag{4}$$

where λ is the mean value of the random variable k. The larger the value, the higher the probability that the technical efficiency will be greatly improved. The technology R&D investment, $R_{i,t}$, is positively correlated with λ, which reflects the second stylized fact of the above mentioned technological progress. The gap between the firm's technical level and the highest technical level, $e_0 - e_{i,t}$, is positively related to λ, which reflects the third stylized fact. $\max_i \{e_{i,t}\} / e_{i,t}$ reflects the gap between the technological level of the firm and the highest technological level in the industry. This value is positively correlated with λ, which reflects the fourth stylized fact of technological progress—the diffusion of advanced technologies in the industry. $\lambda_0, \lambda_1, \lambda_2$, and λ_3 are nonnegative constants.

Finally, when the technical level of the firm reaches its highest boundary value, the firm will stop its R&D investment.

(2) Entry, Progress, and Exit of New Technology

Due to the insufficient supply of feedstock, another stylized fact of China's fuel ethanol firms is that firms need to constantly explore new feedstock and corresponding production technologies. Due to the diversity of fuel ethanol feedstock, the corresponding production technology also shows diverse characteristics. The adoption, progress, diffusion, and withdrawal of different production technologies lead to the change of technological diversity in the industrial technology system, thus promoting the evolution of the technology system.

The entry rules of new technology: This research focuses on the evolution of production technology, which is closely related to industry evolution. Therefore, we use the innovative activities of an R&D firm to describe the evolution of new technology. An R&D firm is a corporation whose output is new technology while the R&D expenditure is its input. We assume that when the industry profit of using traditional technology is negative, new R&D firms start to enter the industry, and the number of entries is random. Among them, the number of R&D firms created by the incumbent firm γ_1 and the number of completely new R&D firms γ_2 is both randomly selected from 0,1......, n_t. Upon entry, all newly created R&D firms are faced with the same initial technical efficiency level, and if an R&D firm already exists in the technology system, the newly created R&D firms will search for the maximum technical efficiency in the existing R&D firms as its initial technical efficiency. The change in the technical efficiency of R&D firms is expressed as follows:

$$\Delta e_{i,t+1} = \varsigma_{i,t} \cdot R_{i,t} \tag{5}$$

In this model, the entry of R&D firms is used to reflect the evolution characteristics of new technologies in the industry. There are two forms of entry for R&D firms: one type of firm is newly created by incumbent firms, and the other is a random start-up. The R&D output of these two kinds of R&D firms is mainly determined by the efficiency of technical output and the amount of R&D input. If the newly established R&D firms have the same total amount of R&D capital, B, and take a fixed proportion of the R&D capital as the R&D investment in each period, the differences between the two kinds of newly established R&D firms include two elements. First, since newly established R&D firms are usually more flexible than incumbent firms, and the flexibility of the system is more conducive to the formation of new technologies, newly established firms will have higher R&D productivity than incumbent firms, which is mainly reflected in the difference between the two types of firms in terms of value $\varsigma_{i,t}$ [43]. Second, as mentioned above, new technology may replace old technology, which will lead to a sunk cost loss for the incumbent firm. Therefore, the incumbent firm will reduce the proportion of R&D investment, which is negatively correlated with the residual fixed assets of the original firm. Its R&D investment is

$$R_{i,t} = t \cdot d \cdot \delta \cdot B \tag{6}$$

and the newly established firms take a fixed amount as the R&D investment:

$$R_{i,t} = \delta \cdot B \tag{7}$$

where δ is a constant. Equations (6) and (7) mean that before the depreciation of fixed assets is completed, the R&D input of the incumbent firm will be less than the R&D input of the new firm.

New technology adoption rules: Firms will only adopt new technology when the average cost of production using that new technology is lower than the average level of traditional technology. This means that, before a firm is able to enter the market, it must go through a long R&D period. During this period, it is difficult for the firm to generate profits and maintain survival. Therefore, in the start-up stage of the new R&D firm, that firm must rely on external capital which is reflected as the initial capital stock of the new R&D firm in this model.

Rules for R&D firms to withdraw from R&D activities: When the initial capital stock is all used for R&D expenditures, and the production costs of new technology still do not reach the average levels of those of traditional technology, the R&D firms will choose to withdraw, which means the withdrawal of new technology.

3.2.3. Market Demand

There are three stylized facts in the production market of the fuel ethanol industry in China. First, fuel ethanol is a typical emerging product whose market demand is gradually forming and expanding or shrinking. Second, there is a competition between fuel ethanol and liquid fossil fuels. The market demand for fuel ethanol comes from the substitution of the market demand for fossil fuels. Third, the government can intervene in the fuel ethanol market through legal or administrative measurements.

In the production market, fuel ethanol is directly incorporated into the formal gasoline sales network, which is an oligopoly market. This means that the price of fuel ethanol will be consistent with gasoline prices, which are decided by the supplier as follows:

$$p_t^b = (1 + \rho_1)p_t^f \tag{8}$$

where p_t^b denotes the price of fuel ethanol, p_t^f denotes the price of gasoline, and ρ_1 denotes the markup percentage.

In pilot provinces, the demand for fuel ethanol will reach a larger scale in the short term after the implementation of the pilot because the pilot consumers can only purchase E10 gasoline. On the other hand, the output of fuel ethanol is relatively smaller in the early stages of the fuel ethanol industry. This means that market demand is horizontal before reaching a specific value but becomes vertical after reaching a specific value. Therefore, Equation (8) can be regarded as the demand function of the fuel ethanol market.

Then, we need to consider the entry, adjustment, or exit of the consumer. As mentioned above, the evolution of the fuel ethanol market is reflected in the entry of potential consumers and the adjustment of consumption by incumbent consumers or the exit from the market.

Consumer Entry Rules: As the consumption of E10 is mandatory, new consumers of fuel ethanol will enter into the market with the expansion of pilot areas.

Consumer Adjustment Rules: When the price of fuel ethanol remains constant, the changes in demand only affected by the changes in consumer's fuel spending. Therefore, consumers will increase or decrease their demand for fuel ethanol based on changes in total fuel spending. The adjustment to fuel ethanol purchases for specific consumers is

$$\dot{x}_{j,t}^b = \dot{M}_{j,t} \tag{9}$$

where $\dot{x}_{j,t}^b$ denotes the percent change in fuel ethanol demand for consumer j in period t; $\dot{M}_{j,t}$ denotes the percent change in consumer j's income in period t.

Consumer Exit Rules: Consumers who are not satisfied with the consumption of fuel ethanol will choose to exit the market.

3.2.4. Industrial Dynamics

(1) Entry of Production Firms

Production firms in the fuel ethanol industry are divided into two types: one is the traditional technology firm which mainly uses traditional production technology and the other is the new technology firm which mainly uses new production technology. Let $\Delta_1 n_t^{tf}$ denote the number of traditional technology firms entering in period t. According to the previous section, it is related to Γ_{t-1}^{tf} and n_{t-1}^{tf}, which indicate the cost to revenue ratio of traditional technology firms and the number of firms in the industry of the last period respectively, as well as the random distribution function p_s. In addition, in the fuel ethanol industry, the entry of traditional technology firms is also restricted by the shortage of feedstocks. Then, we can describe the number of traditional technology firms as follows:

$$\Delta_1 n_t^{tf} = f_1(\Gamma_{t-1}^{tf}, n_{t-1}^{tf}, m_{t-1}^t, p_s) \tag{10}$$

where m_{t-1}^t represents the demand for traditional feedstocks in the last period.

There are two kinds of new technology firms entering in period t in the fuel ethanol industry: one is transformed from R&D firms, which is denoted by $\Delta_1 n_t^{nf}$; the other includes newly established firms that use new technology, denoted by the parameters' values $\Delta_2 n_t^{nf}$. According to the above definition, $\Delta_1 n_t^{nf}$ is related to the number of R&D firms in the last period, n_{t-1}^{rd}, its production costs in the last period, $c_{i,t-1}^{rd}$, and the highest production cost in the last period, $Max\{c_{i,t-1}^t\}$, which can be shown as:

$$\Delta_1 n_t^{nf} = f_2(c_{i,t-1}^{rd}, Max\{c_{i,t-1}^t\}, n_{t-1}^{rd}) \tag{11}$$

$\Delta_2 n_t^{nf}$ is related to the cost of new technology firms in the last period, Γ_{t-1}^{nf}, to the revenue rate, the number of new technology firms in the industry, n_{t-1}^{nf}, and random distribution function, p_s, so:

$$\Delta_2 n_t^{nf} = f_3(\Gamma_{t-1}^{nf},\ n_{t-1}^{nf},\ p_s) \tag{12}$$

(2) Exit of Firms

According to the above rules, when a firm loses money in a continuous k-period, it will exit from production. If the number of withdrawing traditional technology firms in period t is denoted by $\Delta_2 n_t^{tf}$, and the number of withdrawing new technology firms in period t is denoted by $\Delta_3 n_t^{nf}$, then there are

$$\Delta_2 n_t^{tf} = f_4(n_{t-1}^{tf},\ \{\pi_{i,t-1}^{tf},\ \cdots,\ \pi_{i,t-k}^{tf}\}) \tag{13}$$

$$\Delta_3 n_t^{nf} = f_5(n_{t-1}^{nf},\ \{\pi_{i,t-1}^{nf},\ \cdots,\ \pi_{i,t-k}^{nf}\}) \tag{14}$$

(3) Change in the Number of Firms in Production

Then, the number of traditional technology firms in period t is

$$\begin{aligned} n_t^{tf} &= n_{t-1}^{tf} + \Delta_1 n_t^{tf} + \Delta_2 n_t^{tf} \\ &= f_6(n_{t-1}^{tf},\ \Gamma_{t-1}^{tf},\ m_{t-1}^t,\ p_s,\ \{\pi_{i,t-1}^{tf},\ \cdots,\ \pi_{i,t-k}^{tf}\}) \end{aligned} \tag{15}$$

The number of new technology firms in period t is

$$\begin{aligned} n_t^{nf} &= n_{t-1}^{nf} + \Delta_1 n_t^{nf} + \Delta_2 n_t^{nf} + \Delta_3 n_t^{nf} \\ &= f_7(n_{t-1}^{nf},\ c_{i,t-1}^{rd},\ Max\{c_{i,t-1}^t\},\ n_{t-1}^{rd},\ \Gamma_{t-1}^{nf},\ p_s,\ \{\pi_{i,t-1}^{nf},\ \cdots,\ \pi_{i,t-k}^{nf}\}) \end{aligned} \tag{16}$$

The total number of firms involved in the production in the industry at period t is

$$n_t^f = n_t^{tf} + n_t^{nf} = f_8(f_6, f_7) \tag{17}$$

According to Equations (15) and (16), the entry or exit of firms in an industry is influenced not only by the industrial system itself but also by the evolution of the market system and technology system.

4. Results and Discussion

4.1. History Replicating Runs

4.1.1. Baseline Scenario

The aim of the history replicating runs is to find a set of parameters values, based on which the simulation results can reflect the historical stylized facts of the fuel ethanol industry in China [56]. This kind of replicated history is also called the baseline scenario which could be used as a starting point for further policy assessment [57]. Therefore, it is important to select the historical stylized facts that the simulation results intend to capture. Given that the ultimate purpose of our paper is to understand how the fuel ethanol industry has grown, we mainly focused on the following stylized facts about the number of firms and the output of the fuel ethanol industry in China, which are derived from a detailed analysis of the industrial history and from the existing appreciative literature on the fuel ethanol industry in China [58,59].

(1). Once established, traditional technology firms would not withdraw even if they lost money, because the government subsidized loss-making firms to foster the growth of new energy industry.
(2). There is a ceiling of the number of the traditional technology firms because the available raw materials (i.e., expired corn and wheat) used by these firms are limited.

(3).   All R&D firms established by the incumbent traditional technology firms failed to transform into new technology firms.

(4).   All new technology firms are derived from newly established firms with new production technology rather than the firms transforming from R&D firms.

(5).   The number of incumbent firms also has a ceiling because the maximum market capacity is fixed. When the aggregated output of the fuel ethanol beyond the maximum market capacity, there would be no new firms entering the industry.

(6).   The total output increased faster before the number of traditional firms reached the ceiling because the increasement of the total output is mainly caused by the entry of traditional technology firms.

(7).   When the number of traditional technology firms reaches the ceiling, the growth rate of their total output will decrease significantly, because the increasement of the total output is mainly caused by the relatively small expansion of the incumbent firms.

(8).   When the new technology firms start entering the industry, the total output will increase faster again until there is no entry of the new technology firms.

In order to find the set of parameters values which can 'reproduce' the above-mentioned stylized facts, the following steps used in most of the relevant literature were employed in this paper [46,51,60,61]. Firstly, we set the initial values of the variables in the system based on the historical data of the fuel ethanol industry in China. These initial values are shown in Table A1 in Appendix A. Then, we divided the parameters into two groups when setting the values of the parameters. The first group includes the parameters (e.g., the rate of depreciation) which can be set based on the industry history. In these cases, it is possible to fix the parameter to one value. The second group includes other parameters (e.g., the degree of firm rationality) which could not be set to a fixed value for the lack of enough data to estimate its exact value. Thirdly, for the second group of parameters, we did not attempt detailed quantitative matching to historical data, reflecting our ignorance about their 'true' values. Alternatively, we chose values randomly in the predetermined ranges of the parameters and then adjusted the values until the simulation results could qualitatively capture the stylized facts. To adjust the values of the parameters more efficiently, we assessed each parameter's impact on the simulation results, respectively, through keeping all parameters but one constant and then gave preference to adjusting the parameters which had greater impacts. Once the simulation results about the output and the number of firms could qualitatively reflect all the above eight stylized facts, we stopped the adjustment and chose this set of values as the parameters' values under the baseline scenario. Although 'qualitatively reflect' means that there may be many sets of parameters values satisfying the above condition, we did not need to judge which was the best one to reflect these stylized facts because the aim of the history-friendly model was to explore the causal relationships rather than the magnitude of the effects between variables. The final values of parameters are shown in Table A2 in the Appendix A. An additional constraint orienting the choice of parameter values was provided by the time structure of the model, because the definition of what 'one period' means in real-time (half a year in this model) is crucial for establishing which actions take place at any one period. The simulation is implemented using Mathematica software package. The results under the baseline scenario are shown in Figures 1–4.

Figure 1 shows the change in the number of firms in the fuel ethanol industry under the baseline scenario, which can be divided into four stages. There were only traditional technology firms in the first stage, and the number of firms kept increasing. The second stage starts after the 17th period, during which the number of traditional technology firms was no longer increasing. At the same time, R&D firms with new technology appeared, and their numbers increased gradually. However, after the 22nd period, the number of R&D firms began to decline. On the one hand, some R&D firms transformed into production firms and began to produce fuel ethanol. On the other hand, some R&D firms failed to survive because they did not meet the expected innovation goals. The third stage starts in the 26th period. New technology firms began to produce at this stage, and their number increased gradually due to the establishment of the new technology firms and the transformation of R&D firms to new technology firms. According to the simulation results, all production firms transformed from R&D firms are originated from the newly established R&D firms, while the R&D firms created by traditional technology firms failed to transform into production firms, which is consistent with the evolutionary history of the fuel ethanol industry in China. The fourth stage starts from the 31st period. Due to the limitation of market capacity, the number of new technology firms is no longer increasing, and the number of firms in the market remains stable.

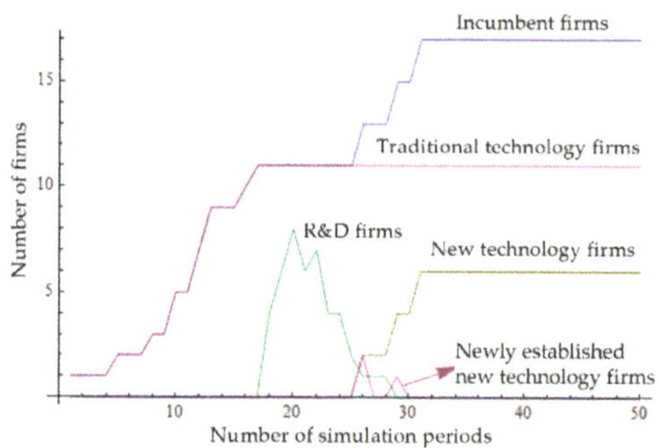

Figure 1. Change in the number of firms.

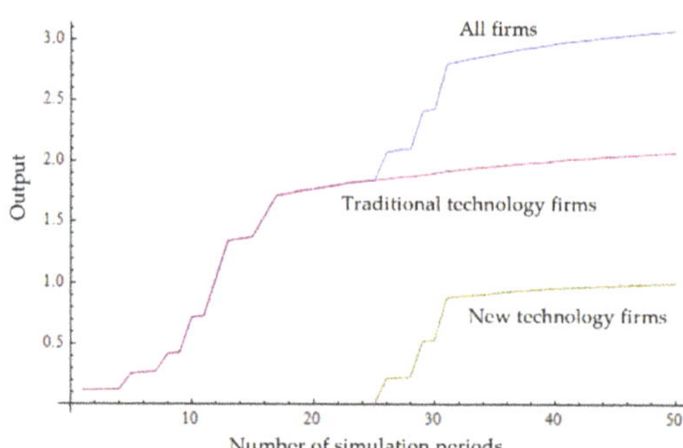

Figure 2. Change in the output of the fuel ethanol industry.

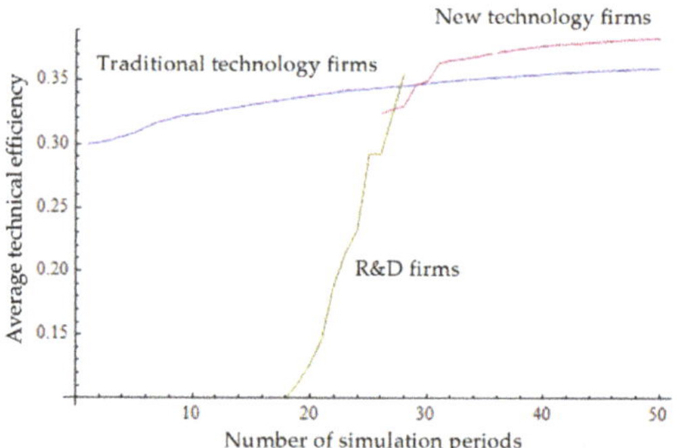

Figure 3. Change of a firm's average technical efficiency.

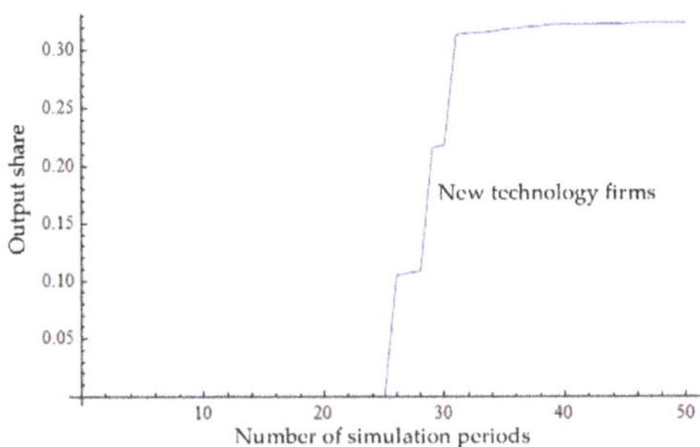

Figure 4. Transformation of the firm's production technology.

Figure 2 shows the change in the output of the fuel ethanol industry under the baseline scenario, which is divided into four stages. In the first stage, the total output of the industry increased continuously. The total output curve coincides with the output of traditional technology firms since there are only traditional technology firms in the industry. The second stage starts from the 17th period when the number of traditional technology firms was not increasing. Although the total output was still increasing, the growth rate of the total output decreased significantly. The increase in output mainly comes from the increase in the output of incumbent firms. The 25th period indicates the third stage: with the entry of new technology firms, the total output began to increase rapidly until the 31st phase. Due to the limitation of market capacity in this period, the number of new technology firms did not increase. The output growth of new technology firms significantly dropped again and then entered the fourth stage. The total output growth is relatively low since there is no entry of new firms. The increase of output in this stage mainly depends on the expansion of the production scale of existing firms.

Figure 3 shows the change of the average technical efficiency of traditional technology firms, R&D firms, and new technology firms under the baseline scenario. The average technical efficiency of all three kinds of firms is increasing. However, the extent of this increase varies from firm to firm. The average technical efficiency of traditional technology firms changes relatively slowly, while the average technical efficiency of R&D firms improves relatively quickly. The average technical efficiency of new technology firms improves rapidly in the initial stage of production but tends to be flat in later stages. The main reason for the above difference is that the technological progress of R&D firms is significantly faster than that of existing production firms. The average technical efficiency of all R&D firms improves relatively quickly because the technical progress among R&D firms is cumulative, and new R&D firms will absorb the experience of existing R&D firms. In the initial stages of production, new technology firms usually adopt the most advanced technology, which significantly improves the average technical efficiency of the whole industry. When the entry of new technology firms stagnates, the improvement of average technical efficiency only depends on the technological progress of incumbent firms. As a result, the rate of technological progress becomes relatively slow.

Figure 4 describes the technology transformation in the fuel ethanol industry under a baseline scenario. Technology transformation refers to the transformation of the industry from traditional production techniques to new production technology. The proportion of the output of new technology firms to the total output of the industry is used to measure technology transformation. According to Figure 4, the value of the technology transformation has rapidly increased since the 25th period and reached about 32% in the 31st period. However, the growth rate of technology transformation has declined rapidly because new technology firms have closed entry to the industry. This means that technology transformation mainly depends on the continuous entry of new technology firms. When there is no continuous entry of new technology firms, the competitive advantage of traditional technology firms in incumbent firms is comparable to that of new technology firms, so, the market share of traditional technology firms will not significantly decrease.

4.1.2. Robustness Test

Simultaneous simulation of "History Replication" and "History Divergent" for the same industry is a commonly used method to test the robustness of the setting of the parameters' values [41,44,62]. The characteristics of the biodiesel industry under the baseline scenario are simulated by adjusting the parameters and setting the initial values according to the stylized facts of the biodiesel industry in China. The simulation results are shown in Figures 5 and 6. The characteristics of the biodiesel industry evolution are significantly different from those of the fuel ethanol industry evolution. This means that the simulation results of the coevolutionary model set in Section 3.2 are indeed affected by the parameter settings. Therefore, the setting of the parameters' values and the initial values under the baseline scenario can be used for further policy analysis.

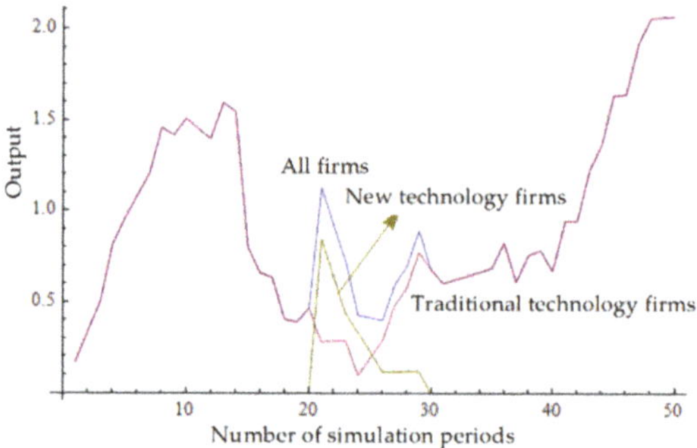

Figure 5. Production change in the biodiesel industry.

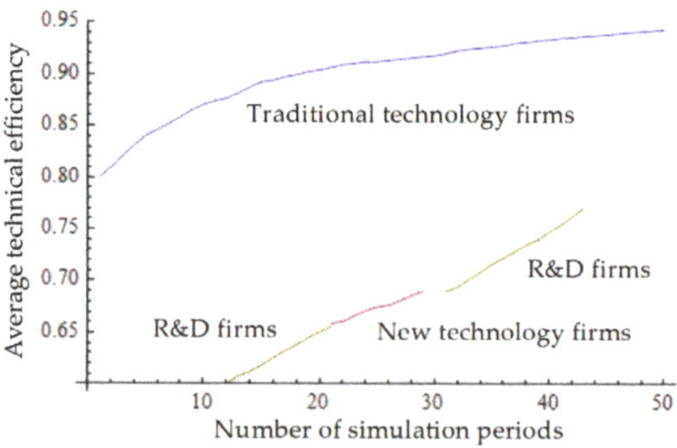

Figure 6. Change of the technical efficiency of the biodiesel industry.

4.2. Policy Impacts Simulation

4.2.1. The Impacts of Entry Regulation

Entry regulation is one of the most important industry regulation policies. The government often uses this kind of policy to avoid overcapacity or other negative social or economic impacts due to overheating industrial development. Entry regulation has also been used in the management of the fuel ethanol industry. How effective is this policy? Does this policy have any other impacts on the industry's growth while achieving its policy objectives? These questions are essential for evaluating the performance of regulation policy. To shed some light on the above questions, we analyzed the impacts of entry regulation on the fuel ethanol industry's evolution using the simulation method. The simulation results are shown in Figures 7–9.

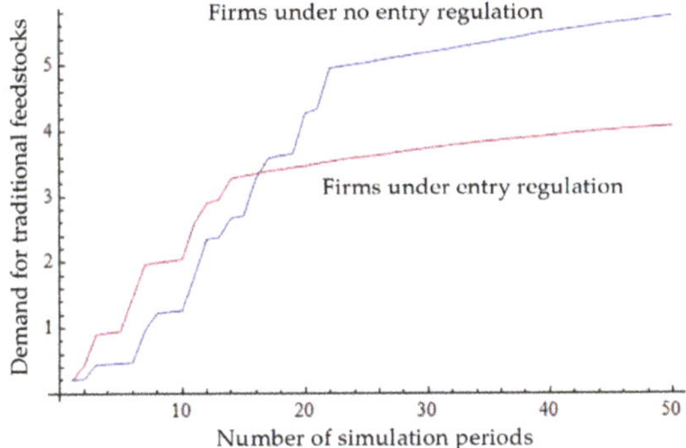

Figure 7. Impact of entry regulation on the demand for feedstock.

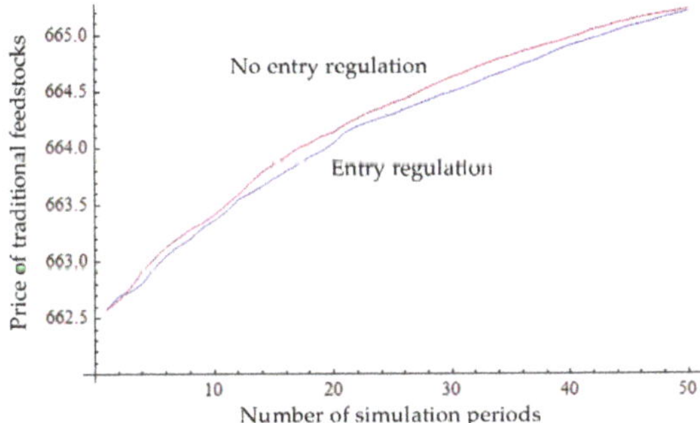

Figure 8. Impact of entry regulation on the price of feedstock.

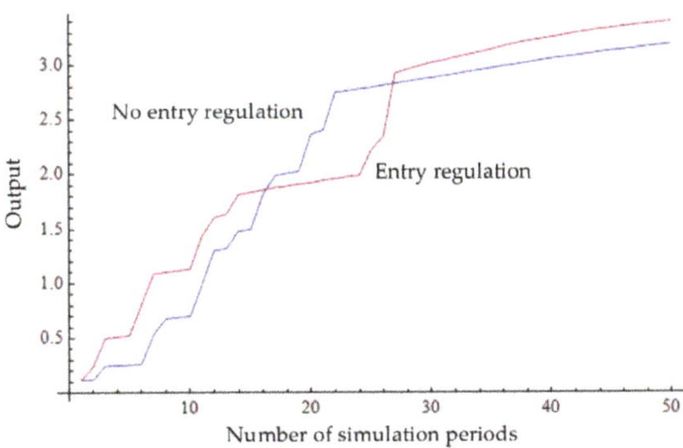

Figure 9. Impacts of entry regulation on fuel ethanol production.

In order to prevent the sharp rise in grain prices due to the rapid development of the traditional grain-based fuel ethanol industry, the Chinese government has implemented an entry regulation that restricts the new establishment of grain-based fuel ethanol firms. The results in Figure 7 show that this regulation policy effectively suppresses the demand for traditional feedstock (e.g., corn and wheat). However, an entry regulation has no significant impact on grain price (shown in Figure 8), because the demand for grain as feedstock for the fuel ethanol industry is only a small part of the total grain demand. Figure 9 shows that the growth of the fuel ethanol production decreases in the short term due to entry regulations. However, in the long run, the restraint effect of entry regulation on production will be eliminated due to the creation of new technology firms. Therefore, entry regulation is conducive to restricting the expansion of the grain-based fuel ethanol industry in the short term, while the long-term impact on the fuel ethanol industry is not too large. Moreover, this policy is helpful in promoting new technology and accelerating technology transformation.

4.2.2. The Impacts of Production Subsidy

A production subsidy is one of the most common policies for the government to promote the development of an emerging industry. Whatever the form of the subsidy, the mechanism of the subsidy is to avoid the losses caused by an immature technology and market in the early stages of the industry so that incumbent firms can continue to produce and improve their technology. With the improvement of technology and the market environment, firms will face competition in the market and make normal profits. Figures 10 and 11 show the impacts of a production subsidy on the output and the number of firms in the fuel ethanol industry, respectively.

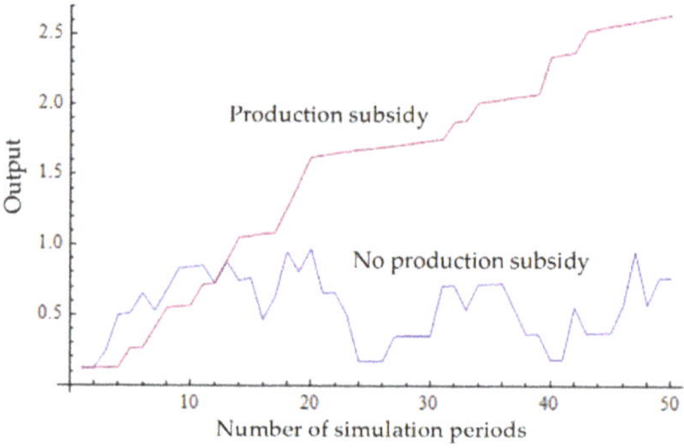

Figure 10. Impact of the production subsidy on the output.

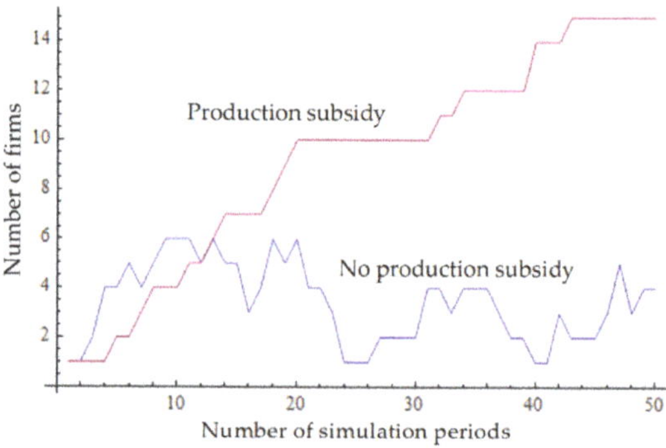

Figure 11. Impact of the production subsidy on the number of firms in the industry.

As shown in Figures 10 and 11, under no production subsidy scenario, firms will continue to enter randomly, and some will withdraw because of losses. Therefore, the industry output and the number of firms will remain at a low level for a long time. However, under the production subsidy scenario, the number of firms entering production will continue to increase, and the output will also continue to rise. The figures also show that the growth rate of the output declined significantly because of the entry regulation on grain-based fuel ethanol firms after the 20th period. Only when new technology firms enter the industry does the output increase significantly again. Therefore, although the simulation results indicate that a subsidy promotes industry growth, this is not a general conclusion. A production subsidy only promotes the growth of subsidized firms and the industries constituted by these firms.

When there are different technological routes in the development of the fuel ethanol industry, the implementation of a single subsidy policy may not promote—or even hinder—the development of the industry.

4.2.3. The Impacts of R&D Subsidy

A subsidy for R&D activities helps to accelerate technological progress by increasing R&D investment. Therefore, an R&D subsidy policy is widely used to promote industry growth. In terms of the fuel ethanol industry, there are not only traditional technology firms but also new technology firms and R&D firms in the industry. This section will analyze the differences in the policy effects of R&D subsidy given to different types of firms. To reduce the influences of random factors, the impacts of the R&D subsidy on average technical efficiency industry output were all simulated 10 times. The simulation results are shown in Figures 12–15.

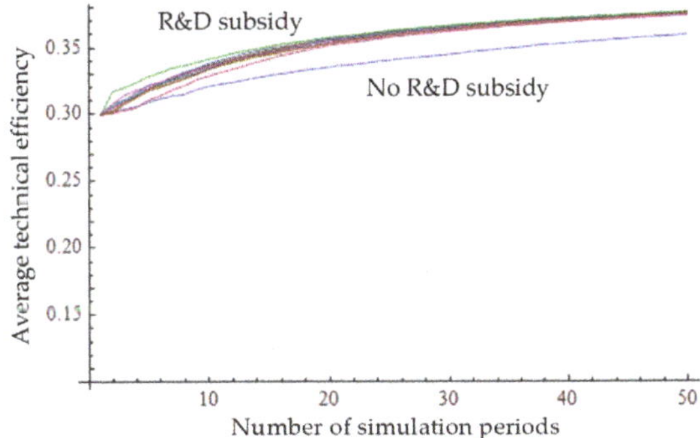

Figure 12. Impact of the R&D subsidy for traditional technology firms on average technical efficiency.

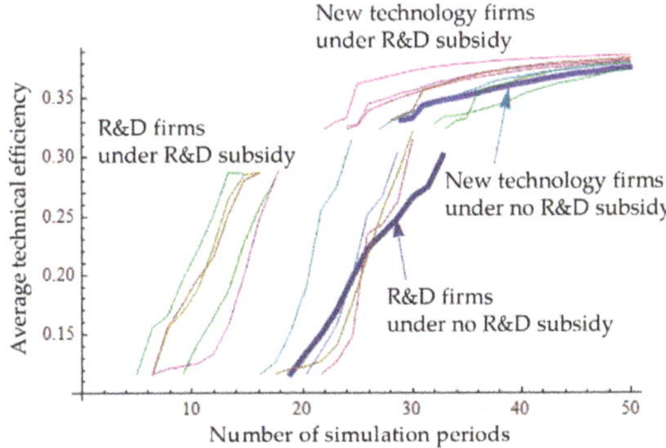

Figure 13. Impact of the R&D subsidy for R&D firms and new technology firms on average technical efficiency.

The results in Figures 12 and 13 show that the R&D subsidy promotes the technical efficiency of the fuel ethanol industry significantly, regardless of the R&D subsidy for traditional technology firms, R&D firms, or new technology firms. However, in terms of industrial scale, the results in Figures 14 and 15 show that the R&D subsidies have no significant impact on industry output, regardless of the subsidy to traditional technology firms or new technology firms. The reason for this result is that firms in the fuel ethanol industry are all in the red and obtain their production subsidy from the government. Therefore, the improvement of firms' technical efficiency, which is partly caused by the R&D subsidy, has little influence on the output decisions of firms. This means that when firms are in the red, the influence mechanism of the R&D subsidy to promote a firm's output will be hindered.

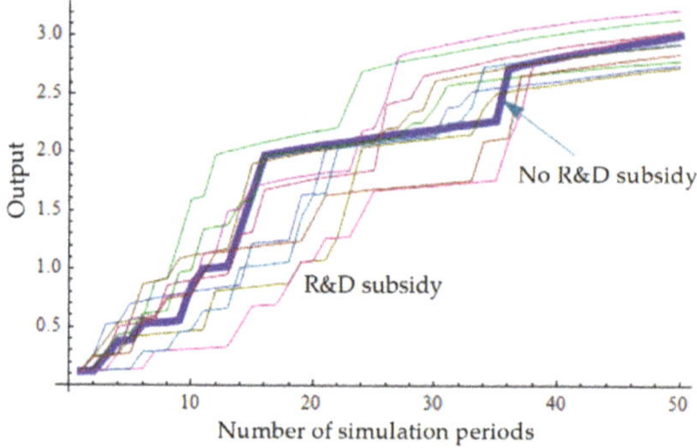

Figure 14. Impact of the R&D subsidy for traditional technology firms on industry output.

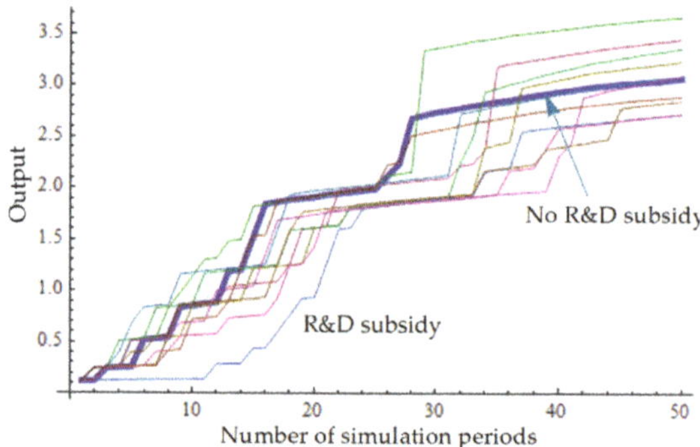

Figure 15. Impact of R&D subsidies for new technology firms on industry output.

4.2.4. The Impacts of Ethanol Mandate

In order to accelerate the development of the fuel ethanol industry in its early stages, the Chinese government implemented an ethanol mandate policy that required the gasoline sold in pilot cities to contain 10% fuel ethanol. Will this policy help the development of the fuel ethanol industry? To answer this question, we simulated the impacts of the ethanol mandate policy on the output of the fuel ethanol industry. The simulation results are shown in Figure 16.

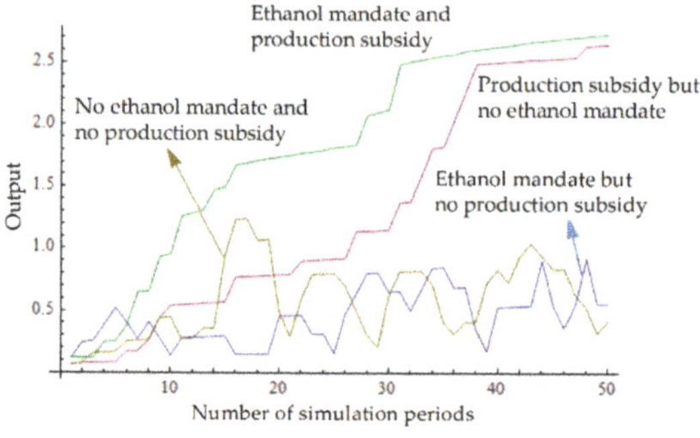

Figure 16. Impact of the ethanol mandate on the output of the fuel ethanol industry.

The result in Figure 16 shows that when there is no production subsidy for losses of firms, the ethanol mandate has no obvious impact on industry output. However, under the subsidy loss scenario, the ethanol mandate will obviously accelerate the development of the fuel ethanol industry. The reason for this phenomenon is that fuel ethanol firms are usually in the red while the industry remains in its embryonic stage. Firms enter the industry randomly and then exit due to continuous losses, so the fuel ethanol industry scale is relatively small. This means that the equilibrium output of the fuel ethanol industry is mainly affected by the number of incumbent firms but not changes in market demand. Therefore, the ethanol mandate has no obvious influence on the output of the fuel ethanol industry, but when firms obtain a loss subsidy, the number of firms will continue to increase, and the industrial output will increase rapidly and reach a larger scale. In this context, the equilibrium output will be mainly affected by market demand. Therefore, an ethanol mandate policy will be helpful to promote the expansion of the fuel ethanol industry.

5. Conclusions and Policy Implications

The interaction among the fuel ethanol industry, the technology system, and the market system has a substantial effect on the growth of the fuel ethanol industry which plays a key role in the formation of a sustainable energy system in China. However, we know little about the relationships among them and it is difficult to explore the nexus using an econometric method due to the lack of statistics on China's fuel ethanol industry. In order to investigate the coevolutionary relationships between the fuel ethanol industry system, technology system, and market system in China, this paper developed a history-friendly simulation model. The setting of the initial values for the simulation model is based on the historical values of the fuel ethanol industry system, technology system, and market system in China. The parameter values of the model were acquired by adjusting the parameters' values continuously until the simulation results could reflect the stylized facts of the fuel ethanol industry. According to the baseline model, this paper further assessed the impacts of entry regulation, production subsidy, R&D subsidy, and ethanol mandate on the growth of the fuel ethanol industry. The results show that multidirectional causalities reflected by the coevolutionary model developed in this paper can describe the relationships between the fuel ethanol industry, technology system, and market system appropriately. This means that the evolution of the fuel ethanol industry interacts with the evolution of the technology system and the market system. Meanwhile, the evolution of the technology system also interacts with the evolution of the market system. Entry regulation is conducive to weakening the negative economic impacts (e.g., rising grain prices and grain shortages) of the expansion of the grain-based fuel ethanol industry without affecting the long-term total output of the industry. A production subsidy for traditional technology firms is helpful to the expansion of the fuel ethanol industry. However, this subsidy also impedes technology transfer in the industry. In terms of R&D subsidy policy, only when the firms inside the industry are not in the red can an R&D subsidy promote technological progress and further accelerate the growth of the fuel ethanol industry. The ethanol mandate policy has a significant impact on industrial expansion only when a production subsidy policy is implemented at the same time. This policy can also speed up the improvement of new technology efficiency by advancing the creation of R&D firms and new technology firms.

According to the above conclusions, the policy suggestions are as follows: firstly, an assessment of the effects of an exogenous factor on any one of these three systems must consider the cumulative effects caused by the coevolutionary mechanisms. Secondly, in the context of this uncertain industry's economic and social impact in China, implementing an entry regulation would be helpful to promote the steady growth of the fuel ethanol industry. Thirdly, when the optimal technology path has not been determined in the Chinese fuel ethanol industry, it will be necessary to implement a production subsidy combined with a new technology promotion policy to avoid technology lock-in. Fourthly, considering that most firms in the fuel ethanol industry in China are still running under a deficit, an R&D subsidy for production firms needs to be implemented along with a production subsidy for unprofitable firms. Lastly, the Chinese government can accelerate the development of the fuel ethanol industry with the help of an ethanol mandate policy. Additionally, the ethanol mandate policy should be implemented together with the production subsidy policy if a firm is running under a deficit.

6. Limitation and Future Research

One limitation of this study is that it does not check the robustness of the simulation results using a statistical approach. It is difficult to estimate the values of the parameters in our model using a statistical method for the lack of enough observations and the complex nonlinearity of the history-friendly model. Alternatively, we employed a process used in most literature about the history-friendly model to determine the values of the parameters in our model. The aim of the process is not the specification of the model parameters as close as possible to their actual values, nor is to explain the quantitative values observed in the historical episode under investigation. Rather, the objective is just to seek a set of parameters' values which can generate simulation results qualitatively capturing the stylized facts of the industrial history, because the purpose of history-friendly modelling is to explore the causal relationships and the mechanism between variables in the model [44]. This means there may be many sets of parameters' values that satisfy the requirement. Therefore, after determining a set of parameters' values for the base case, it is necessary to check the robustness of the simulation results. There are two commonly used approaches to test the robustness in the existing literature [41,47]. One is the inspection of individual runs of the model and the analysis of sensitivity to specific parameter values. The other is the running of a history divergent simulation. These two approaches are also used in this study to check the robustness of the results. However, we suggest there should be a more intense discussion of sensitivity analyses using a statistical method in future research because the above two commonly used approaches could not completely reflect the impacts of stochastic components on the simulation results. Some scholars have already made a bit of progress in this direction. For example, Brenner and Murmann initially developed a process to define the values of the parameters that can be observed with precision [61]. Landini et al. introduced a statistical method which can be used to check the robustness of their results with different variation in the ranges of the parameters [46]. However, both statistical methods have their own drawbacks. The first one turns out not to be immediately comprehensible and could not be completely and clearly described due to the usual limits imposed on the number of words for a paper [50]. The second is still arbitrary in the selection of the variation in the range of parameters. Future research can further improve these two statistical methods and then apply them to the testing of the robustness of the simulation results.

Author Contributions: Conceptualization, C.B. and J.Z.; methodology, C.B. and W.Z.; software, C.B. and Y.W.; formal analysis, C.B. and J.Z.; data curation, W.Z.; writing—original draft preparation, C.B., J.Z., and W.Z; writing—review and editing, C.B.; visualization, C.B. All authors have read and agreed to the published version of the manuscript.

Funding: This research was funded by the fund project of the National Natural Science Foundation of China (Grant No: 71974203), Zhongnan university of economics and law graduate education achievement award cultivation project (Grant No: CGPY201904), interdisciplinary innovation research project (Grant No: 2722019JX002), Zhongnan University of Economics and Law Fundamental Research Funds for the Central Universities (Grant No: 2722020JCT026), the Chinese National Funding of Social Sciences (Grant No: CFA150151), and the Fundamental Research Funds for Shaanxi Normal University (16SZYB34).

Acknowledgments: We thank three anonymous referees for very useful comments.

Conflicts of Interest: The authors declare no conflict of interest.

Appendix A

A1. Modeling the Basic Characteristics of the Firms in the Fuel Ethanol Industry in China

(1) The Production Capacity and Output of the Firm

The production capacity of the firm is related to the fixed capital input of the firm $F_{i,t}$. The variable input of the firm is denoted as

$$m_{i,t} = \eta_{i,t} \cdot (\alpha \cdot F_{i,t}) \tag{A1}$$

where $\eta_{i,t}$ represents the capacity utilization rate of firm i at time t, and $\eta_{i,t} \in [0, 1]$. A value of $\eta_{i,t}$ equal to 0 indicates that the firm stops production. α represents the maximum variable input combined with a one-unit fixed input.

We assume that the output of the firm is determined by the input and technical efficiency. The output of the firm is

$$q_{i,t} = e_{i,t} \cdot (m_{i,t})^z \tag{A2}$$

where $e_{i,t}$ denotes the technical efficiency of firm i at time t, and parameter z ($0 < z < 1$) reflects diminishing marginal productivity.

(2) Production Cost

The total cost of the firm, which consists of the fixed cost and variable cost, is

$$C_{i,t} = d \cdot F_{i,t} + p_t^m \cdot m_{i,t} \tag{A3}$$

where d represents the depreciation rate; $d \cdot F_{i,t}$ denotes fixed cost; p_t^m is the price of feedstock; $m_{i,t}$ is the input amount of the feedstock; $p_t^m \cdot m_{i,t}$ is the variable cost. Considering the relationship between a firm's input and output described by Equation (A2), Equation (A3) can also be expressed as

$$C_{i,t} = d \cdot F_{i,t} + p_t^m \cdot (e_{i,t})^{\frac{-1}{z}} \cdot (q_{i,t})^{\frac{1}{z}} \tag{A4}$$

The cost per unit of the firm is

$$c_{i,t} = C_{i,t} / q_{i,t} \tag{A5}$$

The average per-unit cost of the industry is

$$\bar{c}_{i,t} = \sum_{i=1}^{n_t} C_{i,t} / \sum_{i=1}^{n_t} q_{i,t} \tag{A6}$$

where n_t represents the number of firms in the industry.

(3) Profit of the Firm

Let p_t^b be the price of fuel ethanol; then, the firm's profit is denoted as:

$$\pi_{i,t} = p_t^b \cdot q_{i,t} - C_{i,t} \tag{A7}$$

(4) Supply Function

We assume that a specific firm makes production decisions according to the profit maximization principle; then, the supply function of the firm, which is obtained by taking the derivative of the profit equation (Equation (A7)) with respect to output quantity, is

$$q_{i,t} = (z \cdot p_t^b / p_t^m)^{\frac{z}{1-z}} \cdot e_{i,t}^{\frac{1}{1-z}} \tag{A8}$$

The supply function of the industry, which can be obtained by summing up the supply function of all firms, is

$$q_t = (z \cdot p_t^b / p_t^m)^{\frac{z}{1-z}} \cdot \sum_{i=1}^{n_t} (e_{i,t}^{\frac{1}{1-z}}) \tag{A9}$$

(5) Feedstock demand function

Substitute the relationship between the firm output and the feedstock input of Equation (A2) into Equation (A8); then, the feedstock demand function of the firm can be denoted as

$$m_{i,t} = (z \cdot p_t^b \cdot e_{i,t})^{\frac{1}{1-z}} \cdot (p_t^m)^{\frac{-1}{1-z}} \tag{A10}$$

The feedstock demand function of the industry, which can be obtained by summing up the feedstock demand function of all firms, is

$$m_t = (z \cdot p_t^b)^{\frac{1}{1-z}} \cdot \sum_{i=1}^{n_t} \left(e_{i,t}^{\frac{1}{1-z}} \right) \cdot (p_t^m)^{\frac{-1}{1-z}} \tag{A11}$$

(6) Average Size of the Firm and Technical Efficiency in the Industry

Let the firm's size be denoted by the firm's fixed input, so the average size of the firm in the industry is

$$\overline{F}_t = \sum_{i=1}^{n_t} F_{i,t} / n_t \tag{A12}$$

The average technical efficiency of the industry is reflected by the average technical level of all firms in the industry; then, the average technical efficiency of the industry is

$$\bar{e}_t = \sum_{i=1}^{n_t} e_{i,t} / n_t \tag{A13}$$

The industry's average profit to cost ratio is

$$\Gamma_t = \sum_{i=1}^{n_t} \pi_{i,t} / \sum_{i=1}^{n_t} C_{i,t} \tag{A14}$$

A2. The Initial Values of the Variables in the Coevolutionary Model of the Fuel Ethanol Industry

Table A1. The initial values of the variables in the coevolutionary model of the fuel ethanol industry (baseline scenario).

Initial Values	Description
$n[[1]]=1$	The total number of firms in the first period.
$f[[1,1]][[1]]=0$	Types of technology firm: 0 denotes traditional technology firms; 1 denotes new technology firms.
$f[[1,1]][[2]]=1$	Production status of firms: 0 is no production; 1 is in production.
$f[[1,1]][[3]]=0.3$	Technical efficiency of firms.
$f[[1,1]][[4]]=1000$	Initial capital of firms.
$f[[1,1]][[5]] = 1$	Origin of the production firms: 0 denotes newly established firms; 1 denotes traditional technology firms.
$f[[1,1]][[9]] = 0$	Production status of latecomers: 0 denotes production; 1 denotes stop production.
$pf[[1]] = 2000$	Price of fossil fuel.
$pnm[[1]] = 600$	Price of new feedstock.
$nf[[1]] = 1$	The number of fuel ethanol firms in the first period.
$nnf[[1]] = 0$	The number of fuel ethanol firms using new technology in the first period.
$nntf1[[1]] = 0$	The number of R&D firms established by incumbent fuel ethanol firms in the first period.
$nntf2[[1]] = 0$	The number of newly established R&D firms in the first period.
$nntf[[1]] = 0$	The number of new technology firms in the first period.
$ntf[[1]] = 1$	The number of traditional firms in the first period.

A3. The Setting of the Parameters' Values in the Coevolutionary Model of the Fuel Ethanol Industry

Table A2. The setting of the parameters' values of the fuel ethanol industry (baseline scenario).

Parameter Settings	Description
$\Phi = 0.6$	The degree of firm rationality. Less than 1 denotes imperfect rationality.
$\varphi = 50$	The impact of profit on entry probability.
$d = 0.1$	Rate of depreciation.
$\alpha = 0.21 \times 10^{-3}$	The amount of feedstock matched with the unit capital of traditional technology firm.
$\beta = 0.0002$	The amount of feedstock matched with the unit capital of new technology firm.
$z = 0.6$	Output elasticity of input.
m0 = 3	The maximum supply of feedstock.
efficiency = 0.1	Initial technical efficiency of the first group of R&D firms.
assetn0 = 1000	The initial capital of the R&D firms.
$\delta = 0.2$	The ratio of R&D expenditure to the capital in one period.
totalrd = 20	Total R&D investment of R&D firms.
$\sigma = 0.3$	The proportion of profit used for R&D.
$\varepsilon = 0.3$	Minimum boundary value for capacity utilization.
rdt = 1	Minimum R&D input in each period of the traditional technology firms.
rdn = 2	Minimum R&D input in each period of the incumbent new technology firms.
efftmax = 0.4	The efficiency boundary value of traditional technology firms.
effnmax = 0.4	The efficiency boundary value of new technology firms.
$\lambda 0 = 30$	Parameter related to technological change.
$\lambda 1 - 1$	Parameter related to technological change.
$\lambda 2 = 1$	Parameter related to technological change.
$\lambda 3 = 1$	Parameter related to technological change.
$\kappa 1 = 0.005$	Output efficiency of R&D firms established by incumbent firms.
$\kappa 2 = 0.008$	Output efficiency of newly established R&D firms.
$\rho = 0$	The markup rate of pricing.
b1 = 0	The growth rate of gasoline price
maxdemand = 2.5	The maximum demand in the market.

References

1. Edenhofer, O.; Seyboth, K.; Creutzig, F.; Schlömer, S. On the Sustainability of Renewable Energy Sources. *Annu. Rev. Environ. Resour.* **2013**, *38*, 169–200. [CrossRef]
2. Offermann, R.; Seidenberger, T.; Thrän, D.; Kaltschmitt, M.; Zinoviev, S.; Miertus, S. Assessment of global bioenergy potentials. *Mitig. Adapt. Strat. Glob. Chang.* **2011**, *16*, 103–115. [CrossRef]
3. Hao, M.; Fu, J.; Jiang, D.; Yan, X.; Chen, S.; Ding, F. Sustainable Development of Sweet Sorghum-Based Fuel Ethanol from the Perspective of Water Resources in China. *Sustainability* **2018**, *10*, 3428. [CrossRef]
4. Hu, M.C.; Phillips, F. Technological evolution and interdependence in China's emerging biofuel industry. *Technol. Forecast. Soc. Chang.* **2011**, *78*, 1130–1146. [CrossRef]
5. Chavez, E.; Liu, D.H.; Zhao, X.B. Biofuels Production Development and Prospects in China. *J. Biobased Mater. Bioenergy* **2010**, *4*, 221–242. [CrossRef]
6. Wang, H. Building a regulatory framework for biofuels governance in China: Legislation as the starting point. *Nat. Resour. Forum* **2011**, *35*, 201–212. [CrossRef]
7. Gallagher, P.W. Energy Production with Biomass: What Are the Prospects? *Choices* **2006**, *21*, 21–26.
8. Furtado, A.T.; Scandiffio, M.I.G.; Cortez, L.A.B. The Brazilian sugarcane innovation system. *Energy Policy* **2011**, *39*, 156–166. [CrossRef]
9. Meyer, S.; Binfield, J.; Westhoff, P. Technology adoption under US biofuel policies: Do producers, consumers or taxpayers benefit? *Eur. Rev. Agric. Econ.* **2012**, *39*, 115–136. [CrossRef]
10. Talamini, E.; Dewes, H. The macro-environment for liquid Biofuels in Brazilian science and public policies. *Sci. Public Policy* **2012**, *39*, 13–29. [CrossRef]

11. Tokgoz, S.; Elobeid, A.; Fabiosa, J.; Hayes, D.; Babcock, B.; Yu, T.; Dong, F.; Hart, C.; Beghin, J. Long-Term and Global Trade-offs between Bio-Energy, Feed and Food. Selected Paper Presented at the American Agricultural Economics Association Annual Meeting, Portland, OR, USA, 29 July–1 August 2007.

12. Baker, M.L.; Hayes, D.J.; Babcock, B.A. *Crop-Based Biofuel Production under Acreage Constraints and Uncertainty*; Working Paper 08-WP460; Center for Agricultural and Rural Development, Iowa State University: Ames, IA, USA, 2008.

13. Peters, M.; Stillman, R.; Somwaru, A. Biofuels Expansion in a Changing Economic Environment: A Global Modeling Perspective. In *The Economic Impact of Public Support to Agriculture, Studies in Productivity and Efficiency*; Ball, V.E., Fanfani, R., Gutierrez, L., Eds.; Springer: New York, NY, USA, 2010; Chapter 8; Volume 7, pp. 143–154.

14. Zhang, Y.H.P. What is vital (and not vital) to advance economically-competitive biofuels production. *Process Biochem.* **2011**, *46*, 2091–2110. [CrossRef]

15. Amigun, B.; Sigamoney, R.; von Blottnitz, H. Commercialisation of biofuel industry in Africa: A review. *Renew. Sustain. Energy Rev.* **2008**, *12*, 690–711. [CrossRef]

16. Kang, S.; Önal, H.; Ouyang, Y.; Scheffran, J.; Tursun, Ü.D. Optimizing the Biofuels Infrastructure: Transportation Networks and Biorefinery Locations in Illinois. In *Handbook of Bioenergy Economics and Policy*; Khanna, M., Scheffran, J., Zilberman, D., Eds.; Natural Resource Management and Policy; Springer: New York, NY, USA, 2010; Chapter 10; p. 33.

17. Perkis, D.F.; Tyner, W.E.; Preckel, P.V.; Brechbill, S.C. Spatial Optimization and Economies of Scale for Cellulose to Ethanol Facilities in Indiana. In Proceedings of the Risk, Infrastructure and Industry Evolution. Conference, Berkeley, CA, USA, 24–25 June 2008.

18. Gardner, B. Fuel ethanol subsidies and farm price support. *J. Agric. Food Ind. Organ.* **2007**, *5*, 1–20. [CrossRef]

19. Timilsina, G.R.; Csordás, S.; Mevel, S. When does a carbon tax on fossil fuels stimulate biofuels? *Ecol. Econ.* **2011**, *70*, 2400–2415. [CrossRef]

20. Tyner, W.E. Policy Alternatives for the Future Biofuels Industry. *J. Agric. Food Ind. Organ.* **2007**, *5*, 1123–1156. [CrossRef]

21. Gehlhar, M.; Somwaru, A.; Dixon, P.B. Rimmer, and Ashley, R. Winston. Economywide Implications from US Bioenergy Expansion. *Am. Econ. Rev. Pap. Proc.* **2010**, *100*, 172–177. [CrossRef]

22. Zhang, Z.; Wetzstein, M.E. New Relationships: Ethanol, Corn, and Gasoline Volatility. In Proceedings of the Risk, Infrastructure and Industry Evolution Conference, Berkeley, CA, USA, 24–25 June 2008. [CrossRef]

23. EEA. *How Much Bioenergy can Europe Produce without Harming the Environment?* EEA Report No 7/2006; European Environment Agency; Available online: http://reports.eea.europa.eu/eea_report_2006_7/en (accessed on 8 August 2019).

24. Etter, L. Ethanol Craze Cools As Doubts Multiply. *Wall Str. J.* **2007**. Available online: https://www.wsj.com/articles/SB119621238761706021 (accessed on 10 May 2019).

25. Senauer, B. Food Market Effects of a Global Resource Shift Toward Bioenergy. *Am. J. Agric. Econ.* **2008**, *90*, 1226–1232. [CrossRef]

26. Kallis, G.; Norgaard, R. Coevolutionary ecological economics. *Ecol. Econ.* **2010**, *69*, 690–699. [CrossRef]

27. Fagiolo, G.; Moneta, A.; Windrum, P. A critical guide to empirical validation of agent-based models in economics: Methodologies, procedures, and open problems. *Comput. Econ.* **2007**, *30*, 195–226. [CrossRef]

28. Murmann, J.P. *Knowledge and Competitive Advantage: The Coevolution of Firms, Technology, and National Institutions*; Cambridge University Press: New York, NY, USA, 2003.

29. Goldemberg, J.; Coelho, S.T.; Nastari, P.M.; Lucon, O. Ethanol learning curve-the Brazilian experience. *Biomass Bioenergy* **2004**, *26*, 301–304. [CrossRef]

30. Bastin, C.; Szklo, A.; Rosa, L.P. Diffusion of new automotive technologies for improving energy efficiency in Brazil's light vehicle fleet. *Energy Policy* **2010**, *38*, 3586–3597. [CrossRef]

31. De Souza Nascimento, P.T.; Yu, A.S.O.; Silva, L.L.C.; Starke-Rodrigues, F.C.T.; Morais, C.H.B.; Silva, L.L.; Silva, A.P. The technological strategy of Brazilian automakers for flex-fuel vehicles: An exploratory study. In Proceedings of the PICMET 2010 Technology Management for Global Economic Growth, Phuket, Thailand, 18–22 July 2010; pp. 2846–2858.

32. Sexton, S.E.; Rajagopal, D.; Hochman, G.; Roland-Holst, D.W.; Zilberman, D. Biofuel: Distributional and Other Implications of Current and the Next Generation Technologies. In Proceedings of the Risk, Infrastructure and Industry Evolution Conference, Berkeley, CA, USA, 24–25 June 2008.

33. Perrone, C.C.; Appel, L.G.; MaiaLellis, V.L.; Ferreira, F.M.; De Sousa, A.M.; Ferreira-Leitao, V.S. Ethanol: An evaluation of its scientific and technological development and network of players during the period of 1995 to 2009. *Waste Biomass Valorization* **2011**, *2*, 17–32. [CrossRef]

34. Bigerna, S.; Bollino, C.A.; Micheli, S. Costs assessments of European environmental policies. *Comput. Oper. Res.* **2016**, *66*, 327–335. [CrossRef]

35. Burnes, E.; Wichelns, D.; Hagen, J.W. Economic and policy implications of public support for ethanol production in California's San Joaquin Valley. *Energy Policy* **2005**, *33*, 1155–1167. [CrossRef]

36. Larson, J.; English, B.; He, L. Economic Analysis of Farm-Level Supply of Biomass Feedstocks for Energy Production Under Alternative Contract Scenarios and Risk. In Proceedings of the Transition to a Bio-Economy Conferences, Integration of Agricultural and Energy Systems Conference, Atlanta, GA, USA, 12–13 February 2008.

37. Lasco, C.; Khanna, M. US–Brazil Trade in Biofuels: Determinants, Constraints, and Implications for Trade Policy. In *Handbook of Bioenergy Economics and Policy*; Natural Resource Management and Policy; Khanna, M., Scheffran, J., Zilberman, D., Eds.; Springer: Berlin, Germany, 2010; Chapter 15; p. 33.

38. Sheldon, I.; Roberts, W.U.S. Comparative Advantage in Bioenergy: A Heckscher-Ohlin-Ricardian Approach. *Am. J. Agric. Econ.* **2008**, *90*, 1233–1238. [CrossRef]

39. Thompson, W.; Meyer, S.; Westhoff, P. Policy Risk for the Biofuels Industry. In Proceedings of the Risk, Infrastructure and Industry Evolution Conference, Berkeley, CA, USA, 24–25 June 2008; Burton, C., English, R., Jamey, M., Kim, J., Eds.;

40. Petersen, J.E. Energy production with agricultural biomass: Environmental implications and analytical challenges. *Eur. Rev. Agric. Econ.* **2008**, *35*, 385–408. [CrossRef]

41. Malerba, F.; Nelson, R.R.; Orsenigo, L.; Winter, S.G. 'History-friendly' models of industry evolution: The computer industry. *Ind. Corp. Chang.* **1999**, *8*, 3–40. [CrossRef]

42. Kim, C.W.; Lee, K. Innovation, technological regimes and organizational selection in industry evolution: A 'history friendly model' of the DRAM industry. *Ind. Corp. Chang.* **2003**, *12*, 1195–1221. [CrossRef]

43. Malerba, F.; Orsenigo, L. Innovation and market structure in the dynamics of the pharmaceutical industry and biotechnology: Towards a history-friendly model. *Ind. Corp. Chang.* **2002**, *11*, 667–703. [CrossRef]

44. Malerba, F.; Nelson, R.R.; Orsenigo, L.; Winter, S.G. Vertical integration and disintegration of computer firms: A history-friendly model of the coevolution of the computer and semiconductor industries. *Ind. Corp. Chang.* **2008**, *17*, 197–231. [CrossRef]

45. Brenner, T.; Murmann, J.P. Using simulation experiments to test historical explanations: The development of the German dye industry 1857–1913. *J. Evol. Econ.* **2016**, *26*, 907–932. [CrossRef]

46. Landini, F.; Lee, K.; Malerba, F. A history-friendly model of the successive changes in industrial leadership and the catch-up by latecomers. *Res. Policy* **2017**, *46*, 431–446. [CrossRef]

47. Capone, G.; Malerba, F.; Nelson, R.R.; Orsenigo, L.; Winter, S.G. History friendly models: Retrospective and future perspectives. *Eurasian Bus. Rev.* **2019**, *9*, 1–23. [CrossRef]

48. Nelson, R.R.; Dosi, G.; Helfat, C.E. *Modern Evolutionary Economics: An Overview*; Cambridge University Press: New York, NY, USA, 2018.

49. Malerba, F.; Nelson, R.R.; Orsenigo, L.; Winter, S.G. *Innovation and the Evolution of Industries: History Friendly Models*; Cambridge University Press: Cambridge, UK, 2016.

50. Garavaglia, C. Modelling industrial dynamics with "History-friendly" simulations. *Struct. Chang. Econ. Dyn.* **2010**, *21*, 258–275. [CrossRef]

51. Li, D.; Capone, G.; Malerba, F. The "Long March" to catch-up: A history-friendly model of China's mobile communications industry. *Res. Policy* **2019**, *48*, 649–664. [CrossRef]

52. Tesfatsion, L. Agent-based computational economics: Growing economies from the bottom up. *Artifcial Life* **2002**, *8*, 55–82. [CrossRef]

53. Dawid, H. Agent-based models of innovation and technological change. In *Handbook of Computational Economics*; Agent-Based Computational Economics; Tesfatsion, L., Judd, K., Eds.; Elsevier: Amsterdam, The Netherland, 2006; Volume 2, pp. 1235–1272.

54. Malerba, F.; Nelson, R.; Orsenigo, L.; Winter, S. Competition and industrial policy in a history-friendly model of the evolution of the computer industry. *Int. J. Ind. Organ.* **2001**, *19*, 635–664. [CrossRef]

55. Malerba, F.; Nelson, R.; Orsenigo, L.; Winter, S. Demand, innovation and the dynamics of market structure: The role of experimental users and diverse preferences. *J. Evol. Econ.* **2007**, *17*, 371–400. [CrossRef]

56. Tesfatsion, L. Introduction to the special issue on agent-based computational economics. *J. Econ. Dyn. Control* **2001**, *25*, 281–293. [CrossRef]

57. Winter, S.G.; Kaniovski, Y.M.; Dosi, G. A baseline model of industry evolution. *J. Evol. Econ.* **2003**, *13*, 355–383. [CrossRef]

58. Wu, H.; Li, S. Volatility Spillovers in China's Crude Oil, Corn and Fuel Ethanol Markets. *Energy Policy* **2013**, *62*, 878–886.

59. Jiao, J.; Li, J.; Bai, J. Ethanol as a vehicle fuel in China: A review from the perspectives of raw material resource, vehicle, and infrastructure. *J. Clean. Prod.* **2018**, *180*, 832–845. [CrossRef]

60. Garavaglia, C.; Malerba, F.; Orsenigo, L.; Pezzoni, M. A Simulation Model of the Evolution of the Pharmaceutical Industry: A History-Friendly Model. *J. Artif. Soc. Soc. Simul.* **2013**, *16*, 1–22. [CrossRef]

61. Brenner, T.; Murmann, J.P. The Use of Simulations in Developing Robust Knowledge about Causal Processes: Methodological Considerations and an Application to Industrial Evolution. In *Papers on Economics & Evolution #0303*; Max Planck Institute: Jena, Germany, 2003.

62. Fatas-Villafranca, F.; Jarne, G.; Sanchez-Choliz, J. Industrial leadership in science-based industries: A co-evolution model. *J. Econ. Behav. Organ.* **2009**, *72*, 390–407. [CrossRef]

Article

Techno Economic Evaluation of Cold Energy from Malaysian Liquefied Natural Gas Regasification Terminals

Mohd Amin Abd Majid [1,*], Hamdan Haji Ya [1], Othman Mamat [1] and Shuhaimi Mahadzir [2]

[1] Department of Mechanical Engineering, Universiti Teknologi PETRONAS, Bandar Seri Iskandar 32610, Malaysia; hamdan.ya@utp.edu.my (H.H.Y.); drothman_mamat@utp.edu.my (O.M.)
[2] Department of Chemical Engineering, Universiti Teknologi PETRONAS, Bandar Seri Iskandar 32610, Malaysia; shuham@utp.edu.my
*   Correspondence: mamin_amajid@utp.edu.my

Received: 3 October 2019; Accepted: 20 November 2019; Published: 24 November 2019

Abstract: In order to cater for increased demand for natural gas (NG) by the industry, Malaysia is required to import liquid natural gas (LNG). This is done through PETRONAS GAS Sdn Bhd. For LNG regasification, two regasification terminals have been set up, one in Sungai Udang Melaka (RGTSU) and another at Pengerang Johor (RGTPJ). RGTSU started operation in 2013 while RGTPJ began operation in 2017. The capacities of RGTSU and RGTPJ are 3.8 (500 mmscfd) and 3.5 (490 mmscfd) MTPA, respectively. RGTSU is an offshore plant and uses an intermediate-fluid-vaporization (IFV) process for regasification. RGTPJ is an onshore plant and employs open-rack vaporization (ORV). It is known that a substantial amount of cold energy is released during the regasification process. However, neither plant captures the cold energy released during regasification. This techno economic study serves to evaluate the technical and economic feasibility of the cold energy available during regasification. It was estimated that approximately 47,214 and 88,383 kWh of cold energy could be generated daily at RGTPJ and RGTSU, respectively, during regasification processes. Converting this energy into RTh at 70% thermal efficiency, and taking the commercial rate of 0.549 Sen per RTh, for the 20-year project life, an internal rate of return (IRR) of up to 33% and 17% was estimated for RGTPJ and for RGTSU, respectively.

Keywords: liquefied natural gas; cold energy; regasification; chilled water; techno economic

---

1. Introduction

The share of the liquefied natural gas (LNG) international trade has grown continuously in recent years and LNG has become an important tool for gas security [1]. The traditional supply chain of LNG includes gas production, liquefaction, shipping, storage, and regasification. The practical way to transport natural gas (NG) across oceans is by liquefaction of NG to LNG [2]. This is done by cooling the NG to −162 °C at atmospheric pressure. The LNG is then regasified back to NG at import terminals [3]. Normally, during regasification, the cold energy during the regasification process is discarded. This is also true for LNG regasification terminals in Malaysia. The Malaysian economy grew at 5.51% for the period 2016–2017. Like many developing countries, economic growth has resulted in increased populations in urban areas as well as increased income per capita to catch up with higher living standards, all of which are driving the demand for energy. NG is one of the best choices of primary energy mixes to meet the growing energy demand in modern society due to its clean burning characteristic, high combustion efficiency, and low contribution to greenhouse gases emissions. It is estimated that NG contributes about 24% of Malaysia's energy requirements [4]. To meet the growing demand, Malaysia imports LNG from other producing countries. Currently, two LNG regasification terminals have been built by PETRONAS Gas. The first regasification terminal in Malaysia was set up in Sungai Udang, Melaka (RGTSU), and the second terminal was set up in Pengerang, Johor (RGTPJ). RGTSU started its operation in 2013, and RGTPJ began its operations in the fourth quarter of 2017 [5]. Both terminals are connected to Peninsular Gas Utilization grid pipelines, and then distributed to customers [6].

The RGTSU consists of floating storage and regasification units (FSRUs), and RGTPJ is an onshore terminal. The FSRU is a terminal LNG carrier that has been altered for regasification. Meanwhile, onshore terminals are usually located near the sea. These terminals have operating capacities of 3.8 (500 mmscfd) and 3.5 MTPA (490 mmscfd), respectively [7]. For vaporization, RGTSU employs intermediate fluid vaporization (IFV) technology whereas RGTPJ employs open-rack vaporization (ORV) technology. Table 1 summarizes the information on the terminals.

Table 1. Summary of terminals' information.

No	Item	RGTSU	RGTPJ
1	Facilities		
2	Jetty	Offshore LNGC size:130,000–220,000 m^3 Maximum unloading rate = 10,000 m^3/h	Onshore LNGC size:5000–260,000 m^3 Maximum unloading rate = 14,000 m^3/h
3	Storage	2 units 130,000 m^3 (FSRU)	2 units 200,000 m^3 full containment and LNG tank
4	Vaporization Scheme	IFV with propane as an intermediate fluid and the heating medium is seawater	ORV with sea water as the heating medium
5	Capacity	3.8 MTPA (500 mmscfd)	3.5 MTPA (490 mmscfd)

2. Cold Energy Utilization and Regasification System

2.1. Cold Energy Utilization

He et al. [8] published a review on the current and future utilization of cold energy. A summary of the review is provided in Table 2. In the context of this study, the focus is on the application of waste cold energy for air-conditioning. Waste cold energy from regasification can be captured and stored by using a thermal energy storage (TES) system with chilled water as a cooling medium. The chilled water is used for air conditioning.

Table 2. Current and future utilization of cold energy [8].

	System	Specific Technology	Function of LNG Cold Energy
Current	Cryogenic Power Generation	Organic Rankine Cycle	As heat sink of the cycle
		Brayton Cycle	Reduce the inlet gas temperature
		Kalina Cycle	As heat sink of the cycle
		Combined with gas turbine cycle	Inlet air cooling and intercooling
	Air Separation		Cool the air temperature and replace the external refrigeration cycle
	Seawater Desalination		Cool the seawater
	Cryogenic Carbon Dioxide Capture		Cool and liquefy carbon dioxide

Table 2. *Cont.*

	System	Concepts
Potential	Data Center Cooling	Using LNG cold energy as the source to produce the cooling medium for data center cooling which can reduce energy consumption and greenhouse emissions.
	Clathrate Hydrate-based Desalination	Using LNG cold energy to cool the seawater, hydrate the former, and remove the reaction heat of the clathrate hydrate-based desalination
	Cold Chain for Food Transportation	Using LNG cold energy as the cooling source of the cold warehouse and trucks for storage and ease of transportation
	Cold Energy Storage	Transferring LNG cold energy into an appropriate energy form for longer storage and to ease of transportation
	Utilization of FSRU	Recover LNG cold energy on FSRU by power generation or utilize it for FSRU

2.2. Regasification System

Figure 1 shows the overall process of regasification at RGTSU and RGTPJ.

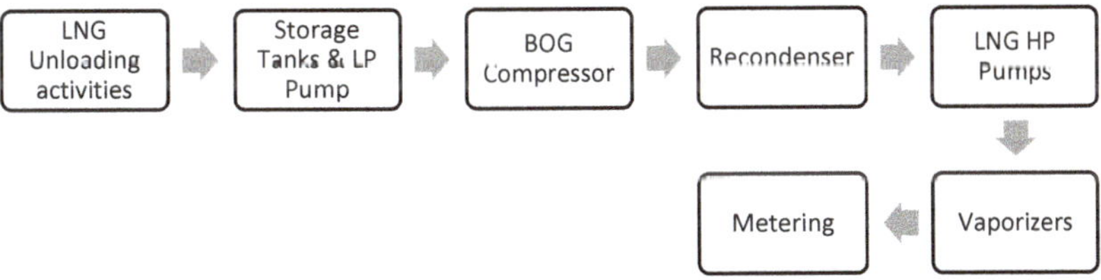

Figure 1. Regasification processes at Regasification Terminal Sungai Udang and Regasification Terminal Pengerang Johor.

During the regasification process, the cold energy of LNG, which is approximately 830 kJ/kg, is released into seawater by LNG vaporizers.

Several vaporization schemes are utilized in regasification technology, including submerged combustion vaporizers (SCRs), ORV, IFV, and super ORVs. A literature survey found that 70% of the regasification terminals used ORV and another 25% and 5% used SCR and IFV, respectively [9,10]. The RGTSU uses IFV (Figure 2). This system consists of two heat exchangers operating in series using propane for intermediate heat transfer (HTF). Propane is used intentionally to prevent the seawater from freezing. The vaporizer is arranged in series to allow the first evaporator exchanger to use the latent heat of propane condensate to partially heat the LNG, and a second heat exchanger uses seawater to further heat the LNG to the required final temperature.

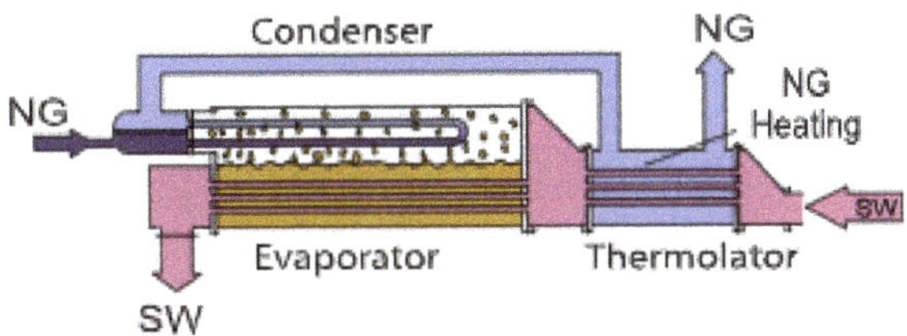

Figure 2. Intermediate-Fluid-Vaporization [11].

The RGTPJ uses ORV (Figure 3). This process employs ribbed-shaped tubes as heat exchangers and seawater as a heat source [11]. The process uses heat transfer between seawater and LNG. Seawater ranging in temperature from 5 to 15 °C is used to heat the LNG from −162 or −163 °C to obtain NG at atmospheric temperature. Seawater temperatures below approximately 5 °C are usually not practical because of seawater freezing [12]. ORV is a well-proven technology and has been widely used in Korea, Europe, and Japanese LNG terminals [10]. Figure 3 shows a schematic diagram of the ORV system.

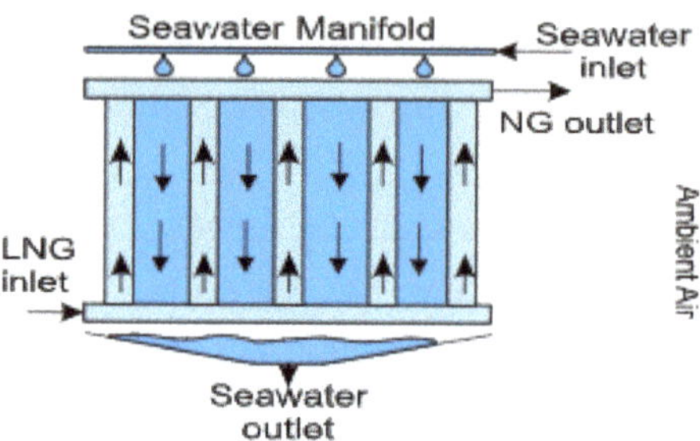

Figure 3. Open-Rack Vaporization [11].

A literature review found that many previous studies exist on LNG cold energy [13–19]. García et al. [20] reported that the regasification of LNG is the last step in the LNG supply chain carried out at LNG terminal storage plants [20]. LNG regasification cold energy can be extensively manipulated into useful energy, which can be applied to cold power generation, seawater desalination, polygeneration, cold air separation, cryogenic crushing, frozen food storage, and carbonic acid production [21]. Such applications can prevent vast stores of cold energy from being thrown away during regasification.

The method for recovering the energy stored in LNG to produce power can be classified into mechanical energy recovery and thermal energy recovery [22]. Mechanical energy recovery uses turbines with LNG as a working fluid [23,24]. Thermal energy recovery uses cycles, such as Rankine, Brayton, and Kalina, and combined forms of these cycles [18,19,25–29]. Despite efforts to utilize this cold energy, approximately 80% of the cold energy from LNG imported globally is still being wasted [30]. The current practice in Malaysia is for cold energy to be released from RGTPJ and RGTSU into the environment via seawater. It is not utilized for any process, whether through mechanical or thermal energy recovery.

Several review papers focused on utilizing LNG cold energy. These reviews mainly focused on progress in power generation utilization without addressing potential applications in which an emerging country, such as Malaysia, can venture. By definition, LNG cold energy utilization systems refer to those requiring low-temperature operating conditions that can be integrated into the LNG regasification process without drastically modifying the system. The potential applications for which cold energy can be utilized without drastically modifying the system include NGL recovery, data center cooling, clathrate hydrate-based desalination, cold chains for food transportation, cold energy storage, and a floating storage regasification unit.

RGTPJ is a land-based regasification terminal that allows for better potential utilization of cold energy for NGL recovery, data center cooling, cold chains for food transportation, and cold energy storage. Meanwhile, RGTSU is best for FSRUs and clathrate hydrate-based desalination given its location offshore. Because of the potential discoveries of these applications, data were gathered from RGTSU and RGTPJ to determine how much cold waste energy can be recovered or potentially utilized through an energy analysis. This study evaluated the potential of using the available cold energy for space cooling by using a TES system.

3. Materials and Method

To evaluate the cold energy available from regasification, temperature, pressure, and flow rate data were acquired for further analysis. These data were acquired at vaporizers and pumps to determine the net energy generated during evaporation. Figure 4 shows the process flow for vaporizers and pumps at RGTSU and Figure 5 shows schematic diagrams of the vaporizers and pumps at RGTPJ. In both flow schemes, LNG from storage at near-atmospheric pressure is sent out through a high-pressure liquid pump to vaporizers. The boiled-off gas from storage is compressed and recondensed before being pumped to the vaporizers.

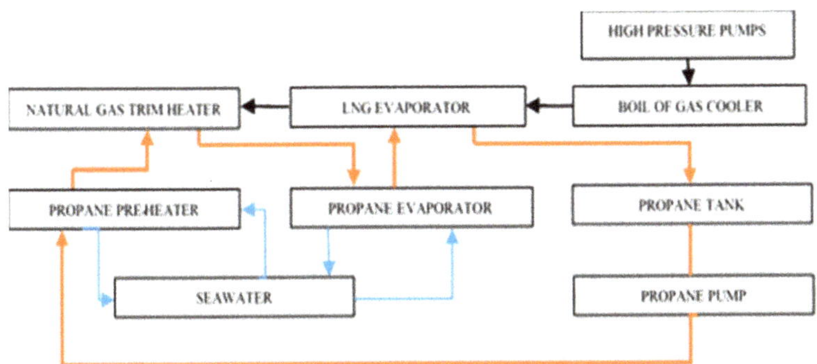

Figure 4. Setup for vaporizers or evaporators and pumps at Regasification Terminal Sungai Udang [31].

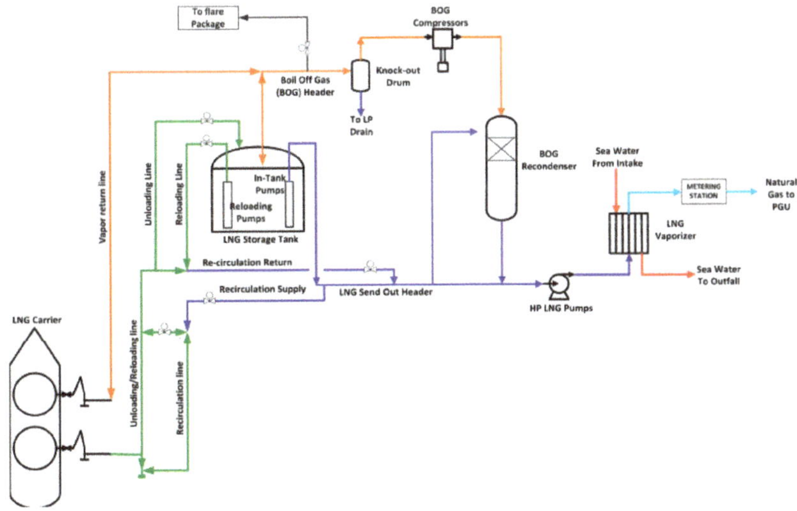

Figure 5. Setup for vaporizers or evaporators and pumps at Regasification Terminal Pengerang Johor [32].

3.1. Energy Models for RGTPJ and RGTSU

RGTPJ and RGTSU regasification processes were simplified as a block diagram, as shown in Figure 6. LNG at −162 °C is heated to normal operating NG between 12 and 20 °C using seawater. With reference to the first law of thermodynamics, the energy balances of the evaporators of RGTPJ and RGTSU were modeled as free-body diagrams, as illustrated in Figure 7. Cold energy generated during the vaporization process of converting LNG to NG is transformed into heat and work energy through the vaporizers.

For this ideal process, the energy available is freely released into the environment. The amount of energy released is estimated using the first law of thermodynamics as per Equation (1):

$$Q = \dot{m}_{sw}\, Cp_{sw}\, (T_{swout} - T_{swin}), \tag{1}$$

where Q, $\dot{m}_{sw}$, Cp_{sw}, T_{swout}, and T_{swin} are the total heat energy (kW), mass flow rate (kg/s), specific heat capacity (kJ/kg °K), and inlet and outlet temperatures (°C) of seawater, respectively.

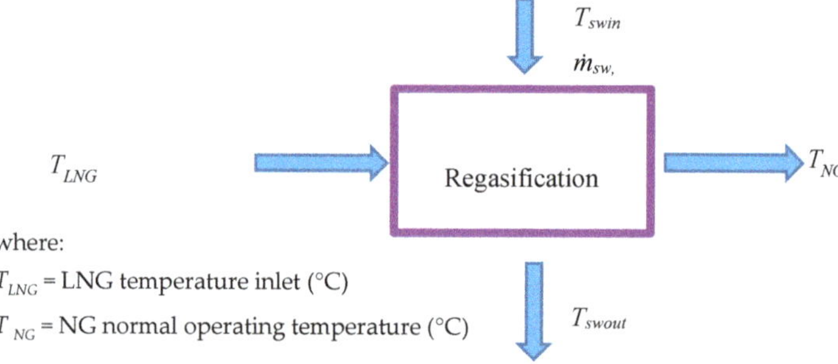

Figure 6. Simplified free-body diagram of the regasification process Regasification Terminal Sungai Udang and Regasification Terminal Pengerang Johor.

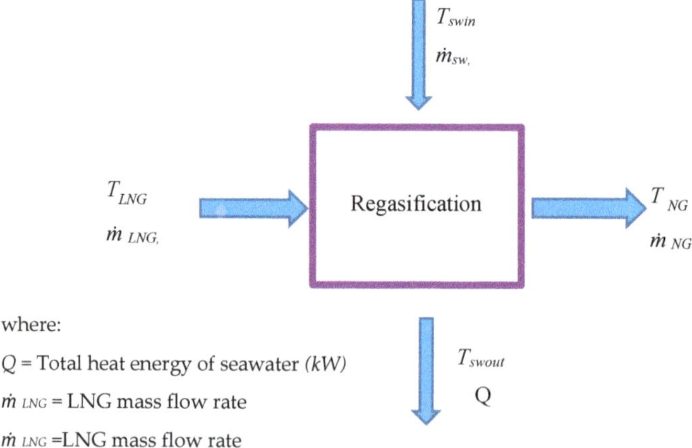

Figure 7. Energy balanced model of regasification Regasification Terminal Sungai Udang and Regasification Terminal Pengerang Johor.

3.2. Economic Models for RGTPJ and RGTSU

To evaluate the economic value of the available cold energy from LNG regasification, an economic analysis was performed. For the analysis, it was assumed that the available waste cold energy is to be converted to the cooling energy of chilled water (CW) at 70% thermal efficiency. A CW system was adopted as the thermal energy storage system (TES) and the CW was used for space cooling. An internal rate of return (IRR) based on the present worth (PW) was adopted for the analysis. IRR was evaluated for a project life from year 1 to year 20 for both RGTPJ and RGTSU. Since the values of the parameters used for evaluating the IRR were based on estimates, sensitivity analyses for IRR were evaluated for years 5, 10, 15, and 20, respectively. Life cycle costing (LCC) for project life of 5, 10, 15, and 20 years were also evaluated for both projects. The steps adopted for the economic analysis are as shown in Figure 8.

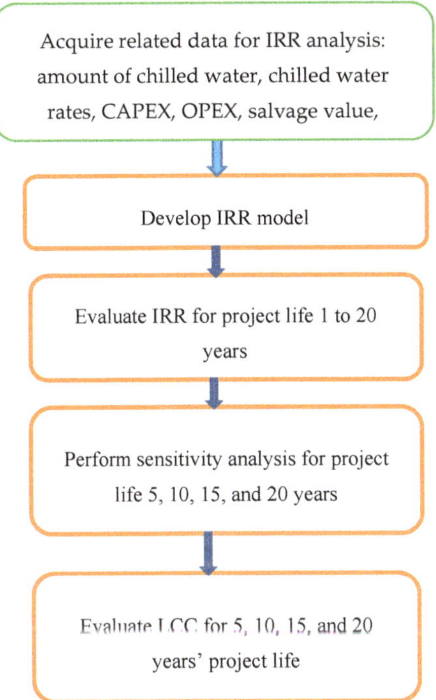

Figure 8. Methodology for the economic analysis for Regasification Terminal Sungai Udang and Regasification Terminal Pengerang Johor.

The steps adopted for the economic analysis were:

i   Identified data required for the analysis. The data included the amount of CW that could be generated from LNG regasification, estimated capital expenses (CAPEXs), operating expenses (OPEXs), salvage value (Sal) of the equipment at the end of the project's life, and the CW rate. The estimated data for both RGTPJ and RGTSU are included in Table 2.

ii   Developed the IRR models for both RGTSU and RGTPJ. The principle used to develop the IRR models was a present worth (PW) analysis of the revenue and the PW of expenses. The PW revenue was taken as being equal to the net annual revenue (NAR), which was equal to the revenue generated from CW minus OPEX as per Equation (2):

$$PW\ of\ NAR + PW\ of\ salvage\ value = CAPEX, \tag{2}$$

$$PW\ NAR = NAR\ (P/A, IRR, N) \tag{3}$$

where:

NAR (nett revenue) = annual revenue – annual OPEX;
NAR (P/A, IRR, N) = PW component for the net revenue;
Sal = salvage value;
Sal (P/F, IRR, N) = PW component for the salvage value;
CAPEX = investment cost;
(P/A, IRR, N) = uniform series PW at discount rate IRR and year N of project life; and
(P/F, IRR, N) = single payment PW at discount rate IRR and year N of project life.

iii   The IRR for RGTPJ and RGTSU were evaluated for project life year 1 up to year 20.

iv   Sensitivity analysis for both RGTPJ and RGTSU was done for the case of project life years 5, 10, 15, and 20 based on the evaluated IRR. Equations (4) and (5) were used for the sensitivity analysis [33]:

$$PW(IRR) = 0 = -CAPEX(1 + x) + NAR\ (P/A, IRR, 5) + 0.02 \times CAPEX\ (P/F, IRR, N),$$
$$N = 5, 10, 15, \text{ and } 20, \tag{4}$$

$$PW(IRR) = 0 == -CAPEX + (NAR\ (1 + y))\ (P/A, IRR, N) + 0.02 \times CAPEX\ (P/F, IRR, N),$$
$$N = 5, 10, 15, \text{ and } 20, \tag{5}$$

where:

For CAPEX, x = percent change in CAPEX; and

For NAR, y = percent change in NAR.

v   Evaluate LCC for the project life of 5, 10, 15, and 20 years.

The LCC models were based on the PW formula for 5, 10, 15, and 20 years. The general LCC model was based on Equation (6).

The LCC models consist of three main components of CAPEX, NAR, and salvage values. The NAR and salvage values were discounted to the current year using the PW formula. The main items that influence the LCC are CAPEX, amount of chilled water, and project life. Hence, if the CAPEX, amount of chilled water and project life change, the IRR will also change, leading to changes in the NAR and salvage value components, and hence, the LCC model:

$$LCC_N = -CAPEX + [RT \times 24 \times 300 \times 0.549 \times 0.8(P/A, IRR_N, N)/1{,}000{,}000 - 0.3 \times CAPEX] +$$
$$0.02 \times CAPEX\ (P/F, IRR_N, N), \tag{6}$$

where the term $+ [RT \times 24 \times 300 \times 0.549 \times 0.8\ (P/A, IRR_N, N)]/1{,}000{,}000 - 0.3 \times CAPEX + 0.02 \times CAPEX\ (P/F, IRR_N, N)$ represents the PW of NAR in RM million discounted to the current with the IRR for the specific N, while the terms $RT \times 24 \times 300 \times 0.549 \times 0.8(P/A, IRR_N, N)$, $0.3 \times CAPEX$, and $0.02 \times CAPEX\ (P/F, IRR_N, N)$ represent the revenue in million RM, annual operating expenses, and the PW of salvage value at the end of year N, respectively.

4. Results and Discussion

4.1. Energy Availability and IRR for the Project Life from Years 1 to 20

The amount of energy availability was calculated using Equation (1) and the following assumptions:

-   No losses on the flow rate of seawater from the evaporation process; and
-   The amount of Q from seawater is 100% converted into energy availability.

Table 3 shows the estimated daily amount of waste cold energy that was available during regasification processes at RGTPJ and RGTSU, respectively. The estimated available waste cold energy daily during regasification are 47,214 and 88,383 kWh at RGTPJ and RGTSU, respectively. In terms of RTh equivalent, the daily amount was 9398 and 17,592 RTh for RGTPJ and RGTSU, respectively. This was based on an assumption of 70% thermal efficiency for the conversion of waste cold energy to chilled water. Using the economic data from Table 4, at 7200 h per year operation, 80% availability, and 0.549RM per RTh, IRR for the project life from year 1 to year 20 were evaluated for RGTPJ and RGTSU. The evaluated IRR for RGTPJ varies from –65% to 33% while for RGTSU, the IRR varies from –80% to 17%. The negative IRR values are IRR during the early years of project life. A Plot of IRR vs. years for both RGTPJ and RGTSU is shown in Figure 9. Results from the IRR analysis indicate that the project could be a profitable venture. RGTPJ gives higher returns compared to RGTSU; the lower returns for RGTSU are due to the higher CAPEX requirements for RGTSU.

Table 3. Energy availability at RGTPJ and RGTSU.

	Seawater Inlet	Seawater Outlet	Energy Availability (kW per hour)	RT/hr (70% Thermal Efficiency Conversion of Energy to CW)
RGTPJ	$m_{fsw} = 5800$ m^3/h $T_{swin} = 30\,°C$	$m_{fsw} = 5800$ m^3/h $T_{swo} = 23\,°C$	47,214	9398
RGTSU	$m_{fsw} = 7600$ m^3/h $T_{swin} = 30\,°C$	$m_{fsw} = 7600$ m^3/h $T_{swo} = 20\,°C$	88,383	17,592

Table 4. Economic analysis assumptions and data for RGTPJ and RGTSU.

	TES Tank Capacity and Auxiliary	Major Equipment Cost/CAPEX (RM)	Annual Expenses (OPEX), RM	Estimated Production Rate and Cost
RGTPJ	• 2 Units: TES Tank @ capacity 10,000 RTh • 2 Units: Heat exchanger @ 250 RT/unit • High pressure pump • Miscellaneous	39.9 M	12.0 M	• CW quantity @ 9398 RT/h • Working hours @ 7200 h/year • CW rates @ RM 0.549/RTh • Availability factor @ 0.8
RGTSU	• 3 Units: TES Tank @ capacity 10,000 RTh • 4 UnitsPlate heat exchanger @ 250 RT/unit • 4 units: High pressure pumps • Miscellaneous	89.34 M	26.8 M	• CW quantity @ 17,592 RT/h • Working hours @ 7200 h/year • CW rates @ RM 0.549/RTh • Availability factor @ 0.8

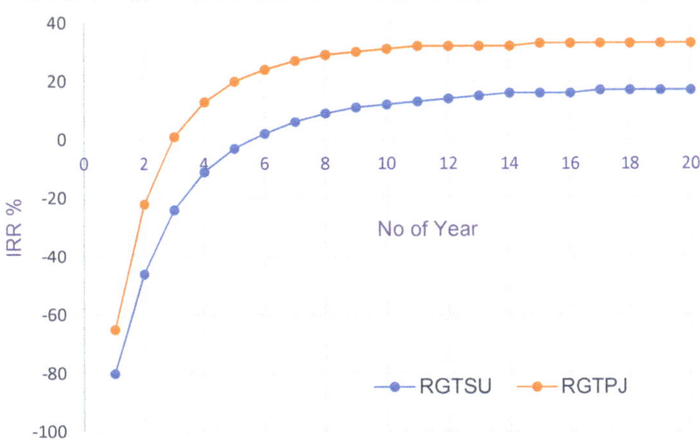

Figure 9. Internal rate of return vs. year for Regasification Terminal Sungai Udang and Regasification Terminal Pengerang Johor.

4.2. Sensitivity Analysis

Since all costs were based on estimates, it is possible that the estimates might not be accurate. It is then essential to evaluate the breakeven points for both CAPEX and NAR. These values were evaluated using Equations (4) and (5), adjusted for RGTSU and RGTPJ as follows:

For RGTSU:

The adjusted sensitivity equation for CAPEX:

$$PW(IRR) = 0 = -89.34(1 + x) + (\text{Annual Revenue - OPEX})_N \ (P/A, IRR, N) + 0.02(89.34)(P/F, IRR, N), N = 5, 10, 15 \text{ and } 20. \tag{7}$$

The adjusted sensitivity equation for net annual revenue (NAR):

$$PW(IRR) = 0 = -89.34 + (\text{Annual revenue - OPEX})_N \ (1 + y)) \ (P/A, IRR, N) + 0.02 \times 89.34(P/F, IRR, N), N = 5, 10, 15 \text{ and } 20. \tag{8}$$

For RTJPJ:

The adjusted sensitivity equation for CAPEX:

$$PW(IRR) = 0 = -39.9(1 + x) + (\text{Annual Revenue - OPEX})_N \ (P/A, IRR, N) + 0.02(39.9)(P/F, IRR, N), N = 5, 10, 15 \text{ and } 20. \tag{9}$$

The adjusted sensitivity equation for net annual revenue (NAR):

$$PW(IRR) = 0 = -39.9 + (\text{Annual revenue - OPEX})_N \ (1 + y)(P/A, IRR, N) + 0.02 \times 39.9(P/F, IRR, N), N = 5, 10, 15 \text{ and } 20. \tag{10}$$

Using Equations (7)–(10) and taking the forecasted annual revenue, OPEX, and the evaluated IRR for the respective project life of 5, 10, 15, and 20 years, the values of x and y were calculated for both RGTPJ and RGTSU, respectively. The evaluated results are included in Table 5.

Table 5. Sensitivity of Internal rate of return for Capital expenditure and Nett annual revenue

RGTPJ	IRR	x (CAPEX)	y (Nett Revenue)	Remarks
5	20	0.34	−1.75	If CAPEX increases by more than 0.34% project not viable If NAR is lower by more than 1.75% project not viable
10	31	0.34	−1.75	If CAPEX increases by more than 0.34% project not viable If NAR is lower by more than 1.75% project not viable
15	33	0.33	−1.75	If CAPEX increases by more than 0.34% project not viable If NAR is lower by more than 1.75% project not viable
20	33	0.34	−1.74	If CAPEX increases by more than 0.34% project not viable If NAR is lower by more than 1.74% project not viable
RGTSU	**IRR %**	**x**	**y**	**Remarks**
5	−3	NA	−NA	NA due to negative IRR
10	12	0.83	−1.54	If CAPEX increases by more than 0.83% project not viableIf NAR is lower by more than 1.54% project not viable
15	16	0.8	−1.55	If CAPEX increases by more than 0.8% project not viableIf NAR is lower by more than 1.55% project not viable
20	17	0.82	−1.55	If CAPEX increases by more than 0.82% project not viableIf NAR is lower by more than 1.55% project not viable

For RGTPJ, the sensitivities for CAPEX are 0.34%, 0.34%, 0.34%, and 0.34% for years 5, 10, 15, and 20, respectively. For NAR, the sensitivities are −1.7%5, −1.75%, −1.75%, and −1.74% for years 5, 10, 15, and 20, respectively.

For RGTSU, the CAPEX sensitivities are NA, 0.83%, 0.8%, and 0.82% for years 5, 10, 15, and 20, respectively. While for NAR, sensitivities are NA, −1.54%, −1.55%, and −1.55% for years 5, 10, 15, and 20, respectively.

4.3. LCC Models

Using Equation (6), the LCC models for RGTPJ and RGTSU were formulated. The LCC models are included in Table 6.

Table 6. Life cycle cost models for Regasification Terminal Sungai Udang and Regasification Terminal Pengerang Johor for year 5, 10, 15, and 20.

Project Life	RGTPJ $LCC_N = -CAPEX + [RT \times 24 \times 300 \times 0.549 \times 0.8(P/A, IRR_N, N)/1{,}000{,}000 - 0.3 \times CAPEX] + 0.02 \times CAPEX(P/F, IRR_N, N)$
5	$LCC_5 = -39.9 + [9398 \times 24 \times 300 \times 0.549 \times 0.8(P/A, 20{,}5)/1{,}000000 - 0.3 \times 39.9] + 0.02 \times 39.9(P/F, 20{,}5)$
10	$LCC_{10} = -39.9 + [9398 \times 24 \times 300 \times 0.549 \times 0.8(P/A,31{,}10)/1{,}000{,}000 - 0.3 \times 39.9] + 0.02 \times 39.9(P/F, 31{,}10)$
15	$LCC_{15} = -39.39 + [9398 \times 24 \times 300 \times 0.549 \times 0.8(P/A, 33{,}15)/1{,}000{,}000 - 0.3 \times 39.9] + 0.02 \times 39.9(P/F, 33{,}15)$
20	$LCC_{20} = -39.9 + [9398 \times 24 \times 300 \times 0.549 \times 0.8(P/A, 33{,}20)/1{,}000{,}000 - 0.3 \times 39.9] + 0.02 \times 39.9(P/F, 33{,}20)$
	RGTSU$LCC_N = -CAPEX + [RT \times 24 \times 300 \times 0.549 \times 0.8(P/A, IRR_N, N)/1{,}000{,}000 - 0.3 \times CAPEX] + 0.02 \times CAPEX (P/F, IRR_N, N)$
5	$LCC_5 = -89.34 + [17592 \times 24 \times 300 \times 0.549 \times 0.8(P/A, -3{,}5)/1{,}000{,}000 - 0.3 \times 89.34] + 0.02 \times 89.34(P/F, -3{,}5)$
10	$LCC_{10} = -89.34 + [17592 \times 24 \times 300 \times 0.549 \times 0.8(P/A, 12{,}10)/1{,}000{,}000 - 0.3 \times 89.34] + 0.02 \times 89.34(P/F,12{,}10)$
15	$LCC_{15} = -89.34 + [17592 \times 24 \times 300 \times 0.549 \times 0.8(P/A, 16{,}15)/1{,}000{,}000 - 0.3 \times 89.34] + 0.02 \times 89.34(P/F, 16{,}15)$
20	$LCC_{20} = -89.34 + [17592 \times 24 \times 300 \times 0.549 \times 0.8(P/A, 17{,}20)/1{,}000{,}000 - 0.3 \times 89.34] + 0.02 \times 89.34(P/F, 17{,}20)$

It is noted that the CAPEX, the present worth components for the revenue from chilled water, operating cost, and to lesser extent, the salvage value influence the LCC.

Using the equations in Table 6, LCC for RGTSU and RGTPJ were evaluated. Results for both RGTSU and RGTPJ are tabulated in Table 7.

Table 7. Evaluated Life cycle cost for Regasification Terminal Sungai Udang and Regasification Terminal Pengerang Johor.

RGTSU		
Project Life	**IRR**	**LCC Value (million RM)**
5	−3	NA due to negative IRR
10	12	198.74
15	16	194.22
20	17	197.01
RGGPJ		
Project Life	**IRR**	**LCC Value (million RM)**
5	20	37.33
10	31	39.61
15	33	36.95
20	33	37.88

The LCC results for RGTSU vary from 197 to RM198.7 which are of higher values compared to RGTPJ LCC values, which vary from RM37.3 million to RM39.6 million. This is due to higher CAPEX value for RGTSU compared to RGTPJ.

5. Conclusions

Currently, the two regasification terminals operated by PETRONAS Gas Sdn Bhd do not capture waste cold energy during the regasification process. This study noted that substantial waste cold energy is available during regasification at both RGTPJ and RGTSU. The estimated annual amount of cold energy that could be captured daily during regasification at RGTPJ and RGTSU is 47,214 and 88,383 kWh, respectively. The study evaluated the commercial potential of using the available cold energy for chilled water generation. The chilled water is to be used for space cooling. Assuming 70% thermal efficiency conversion of waste cold energy to chilled water, it was estimated that daily, the amount of cold energy available hourly during regasification at RGTPJ and RGTSU is equivalent to 9398 and 17,592 RTh amount of chilled water, respectively. From the economic feasibility study, commercially, the revenue from the chilled water could give IRR greater than 20% for RGTPJ for a project life of 5 to 20 years. For RGTSU, the IRR values are 12% to 17% for a project life of 10 to 20 years. Hence, if the waste cold energy during regasification at RGTPJ and RGTSU is exploited, it would give a profitable venture. In addition, the venture would also increase the efficiency of LNG regasification at both terminals and the economic benefit of the LNG supply chain. Besides using the cold energy for generating CW, the waste cold energy could also be used to cool intake air for the gas turbines. Since RGTPJ is located near the vicinity of a cogeneration plant, the cold energy from regasification should also be considered for use for cooling intake air for the gas turbines at the cogeneration plant. It is therefore recommended that the owners of RGTPJ and RGTSU should consider installing systems able to capture the waste cold energy during regasification of LNG at both terminals.

Author Contributions: Conceptualization, M.A.A.M. and H.H.Y.; Methodology, M.A.A.M. and H.H.Y.; Software, O.M.; Validation, S.M.; Formal Analysis, M.A.A.M. and H.H.Y.; Investigation, O.M.; Resources, O.M.; Data Curation, S.M.; Writing-Original Draft Preparation, M.A.A.M. and H.H.Y.; Writing-Review & Editing, M.A.A.M. and H.H.Y.; Visualization, O.M.; Supervision, M.A.A.M.; Project Administration, M.A.A.M.; Funding Acquisition, M.A.A.M. and O.M.

Acknowledgments: This work represents collaborative research among Universiti Teknologi PETRONAS, Malaysia, and Heriot-Watt University, U.K., and was funded by the British Council's Newton Institutional Links entitled "Building institutional links to deliver sustainable cooling energy demand." The authors would like to acknowledge the support of RGTPJ and RGTSU for this research.

Conflicts of Interest: The authors declare no conflict of interest.

Nomenclature

RGTSU	Regasification terminal Sungai Udang Melaka
RGTPJ	Regasification terminal Pengerang Johor
LNG	Liquified natural gas
LNGC	LNG Carrier
NG	Natural gas
CW	Chilled water
Q	Total waste cold energy
$\dot{m}_{sw}$	Seawater mass flow rate
Cp_{sw}	Seawater specific heat capacity
T_{swout}	Seawater temperature outlet
T_{swin}	Seawater temperature inlet
RTh	Refrigeration ton hour
TES	Thermal energy storage
CAPEX	Capital cost
OPEX	Operation expenses
NAR	Nett annual revenue
PW	Present worth
Sal	Salvage value
IRR	Internal rate of return
LCC	Life cycle costing

References

1. Sun, Z.; Xu, F.; Wang, S.; Lai, J.; Lin, K. Comparative study of Rankine cycle configurations utilizing LNG cold energy under different NG distribution pressures. *Energy* **2017**, *139*, 380–393. [CrossRef]
2. Sun, H.; Zhu, H.; Liu, F.; Ding, H. Simulation and optimization of a novel Rankine power cycle for recovering cold energy from liquefied natural gas using a mixed working fluid. *Energy* **2014**, *70*, 317–324. [CrossRef]
3. Gómez, M.R.; Garcia, R.F.; Gómez, J.R.; Carril, J.C. Thermodynamic analysis of a Brayton cycle and Rankine cycle arranged in series exploiting the cold exergy of LNG (liquefied natural gas). *Energy* **2014**, *66*, 927–937. [CrossRef]
4. Malaysia Energy Commission. Malaysia Energy Information Hub. 2012. Available online: http://meih.st.gov.my/statistics (accessed on 14 April 2015).
5. Ezhar, Y.J. *Challenges and Prospects in Enhancing Malaysia's Energy Security*; PETRONAS: Kuala Lumpur, Malaysia, 2012.
6. Berhad, P.N. *Petronas Annual Report 2011*; Petroliam Nasional Berhad: Kuala Lumpur, Malaysia, 2012.
7. Bujang, A.; Bern, C.; Brumm, T. Summary of energy demand and renewable energy policies in Malaysia. *Renew. Sustain. Energy Rev.* **2016**, *53*, 1459–1467. [CrossRef]
8. He, T.; Chong, Z.R.; Zheng, J.; Ju, Y.; Linga, P. LNG cold energy utilisation: Prospects and challenges. *Energy* **2019**, *170*, 557–568. [CrossRef]
9. Vatani, A.; Mehrpooya, M.; Palizdar, A. Energy and exergy analyses of five conventional liquefied natural gas processes. *Int. J. Energy Res.* **2014**, *38*, 1843–1863. [CrossRef]
10. Patel, D.; Mak, J.; Rivera, D.; Angtuaco, J. LNG vaporizer selection based on site ambient conditions. *Proc. LNG* **2013**, *17*, 16–19.
11. Hisada, N.; Sekiguchi, M. Design and analysis of open rack LNG vaporizer. In Proceedings of the ASME/JSME Pressure Vessels and Piping Conference, San Diego, CA, USA, 25–29 July 2004; pp. 97–104.
12. Tarlowski, J.; Sheffield, J.; Durr, C.; Coyle, D.; Patel, H. *LNG Import Terminals-Recent Developments*; MW Kellogg Ltd.: London, UK, 2005.
13. Castillo, L.; Dorao, C. On the conceptual design of pre-cooling stage of LNG plants using propane or an ethane/propane mixture. *Energy Convers. Manag.* **2013**, *65*, 140–146. [CrossRef]
14. Alabdulkarem, A.; Mortazavi, A.; Hwang, Y.; Radermacher, R.; Rogers, P. Optimization of propane pre-cooled mixed refrigerant LNG plant. *Appl. Therm. Eng.* **2011**, *31*, 1091–1098. [CrossRef]
15. Hatcher, P.; Khalilpour, R.; Abbas, A. Optimisation of LNG mixed-refrigerant processes considering operation and design objectives. *Comput. Chem. Eng.* **2012**, *41*, 123–133. [CrossRef]
16. Gavelli, F. Computational fluid dynamics simulation of fog clouds due to ambient air vaporizers. *J. Loss Prev. Process Ind.* **2010**, *23*, 773–780. [CrossRef]
17. Liu, Y.; Guo, K. A novel cryogenic power cycle for LNG cold energy recovery. *Energy* **2011**, *36*, 2828–2833. [CrossRef]
18. Lu, T.; Wang, K. Analysis and optimization of a cascading power cycle with liquefied natural gas (LNG) cold energy recovery. *Appl. Therm. Eng.* **2009**, *29*, 1478–1484. [CrossRef]
19. Shi, X.; Che, D. A combined power cycle utilizing low-temperature waste heat and LNG cold energy. *Energy Convers. Manag.* **2009**, *50*, 567–575. [CrossRef]
20. García, R.F.; Carril, J.C.; Gomez, J.R.; Gomez, M.R. Combined cascaded Rankine and direct expander based power units using LNG (liquefied natural gas) cold as heat sink in LNG regasification. *Energy* **2016**, *105*, 16–24. [CrossRef]
21. Atienza-Márquez, A.; Bruno, J.C.; Coronas, A. Cold recovery from LNG-regasification for polygeneration applications. *Appl. Therm. Eng.* **2018**, *132*, 463–478. [CrossRef]
22. Lee, S. Multi-parameter optimization of cold energy recovery in cascade Rankine cycle for LNG regasification using genetic algorithm. *Energy* **2017**, *118*, 776–782. [CrossRef]
23. Qiang, W.; Yanzhong, L.; Jiang, W. Analysis of power cycle based on cold energy of liquefied natural gas and low-grade heat source. *Appl. Therm. Eng.* **2004**, *24*, 539–548. [CrossRef]
24. Franco, A.; Casarosa, C. Thermodynamic analysis of direct expansion configurations for electricity production by LNG cold energy recovery. *Appl. Therm. Eng.* **2015**, *78*, 649–657. [CrossRef]
25. Hisazumi, Y.; Yamasaki, Y.; Sugiyama, S. Proposal for a high efficiency LNG power-generation system utilizing waste heat from the combined cycle. *Appl. Energy* **1998**, *60*, 169–182. [CrossRef]

26. Zhang, N.; Lior, N. A novel near-zero CO2 emission thermal cycle with LNG cryogenic exergy utilization. *Energy* **2006**, *31*, 1666–1679. [CrossRef]

27. Wang, J.; Yan, Z.; Wang, M.; Dai, Y. Thermodynamic analysis and optimization of an ammonia-water power system with LNG (liquefied natural gas) as its heat sink. *Energy* **2013**, *50*, 513–522. [CrossRef]

28. Deng, S.; Jin, H.; Cai, R.; Lin, R. Novel cogeneration power system with liquefied natural gas (LNG) cryogenic exergy utilization. *Energy* **2004**, *29*, 497–512. [CrossRef]

29. Gómez, M.R.; Garcia, R.F.; Gómez, J.R.; Carril, J.C. Review of thermal cycles exploiting the exergy of liquefied natural gas in the regasification process. *Renew. Sustain. Energy Rev.* **2014**, *38*, 781–795. [CrossRef]

30. Agarwal, R.; Babaie, M. LNG regasification—Technology evaluation and cold energy utilisation. *IGT Int. Liq. Nat. Gas Conf. Proc.* **2013**, *3*, 2134–2142.

31. Regasification Terminal Sungai Udang. *Accsee Arrangement for Regasification Terminal Sungai Udang*; PETRONAS: Kuala Lumpur, Malaysia, 2018.

32. Pengerang LNG (Two) Sdn. Bhd. *Access Arrangement for Regasification Terminal Pengerang*; PETRONAS: Kuala Lumpur, Malaysia, 2018; p. 121.

33. Sullivan, W.G.; Wicks, E.M.; Koelling, C.P. *Engineering Economy*, 16th ed.; Prentice Hall International: New York, NY, USA, 2014.

Article

Interdependence Between Renewable-Energy and Low-Carbon Stock Prices

Juan C. Reboredo [1,*,†], **Andrea Ugolini** [2,‡] **and Yifei Chen** [1,‡]

[1] Department of Economics, Universidade de Santiago de Compostela, 15782 Santiago de Compostela, Spain; chenyifei9@gmail.com

[2] Departament of Quantitative Analysis, Universidade do Estado do Rio de Janeiro, 20550-900 Rio de Janeiro, Brazil; andreaugolini@me.com

[*] Correspondence: juancarlos.reboredo@usc.es; Tel.: +34-881-811-675

[†] Current address: Departamento de Fundamentos del Análisis Económico, Universidade de Santiago de Compostela, Avda. Xoán XXIII, s/n, 15782 Santiago de Compostela, Spain.

[‡] These authors contributed equally to this work.

Received: 22 October 2019; Accepted: 18 November 2019; Published: 22 November 2019

Abstract: In the transition to a low-carbon economy, climate-resilient investors may be inclined to buy renewable-energy or other low-carbon assets. As the diversification benefits of investment positions in those assets depend on interdependence between their market prices, we explore that interdependence in the European and USA stock markets. We model the dependence structure using bivariate copula functions and evaluate price spillovers between those markets using a conditional quantile dependence approach that accounts for the reciprocal effects of price movements in those markets under normal and extreme market scenarios. Our empirical evidence for the period 2010–2019 indicates that European renewable-energy and low-carbon stocks co-move; upward and downward movements in low-carbon asset prices have sizeable effects on renewable-energy asset prices, and vice versa, although effects are smaller. In contrast, for the USA we find evidence of non-interdependence, with no significant upward or downward price spillover effects between renewable-energy and low-carbon stocks. Our empirical findings provide useful insights for the design of carbon-resilient portfolios and risk management strategies, and also for implementation of public funding policies to support the transition to a low-carbon economy.

Keywords: renewables; low carbon; interdependence; copulas; conditional quantiles

JEL Classification: C22; C58; F30; G11; G15

---

1. Introduction

The transition to a low-carbon economy entails a vast amount of financial resources, which, in turn, raises awareness among investors about opportunities and risks linked to that transition. Renewable-energy and low-carbon assets are arguably the most suitable investment vehicles to ensure private capital reallocation that meets the challenges posed by decarbonization. Therefore, understanding interdependence between the prices of renewable-energy and low-carbon assets is essential information for environmentally-friendly investors, as it determines the diversification benefits of allocating private capital to climate-resilient portfolios and shapes private incentives to deploy financial resources to clean energies and low-carbon industries. Moreover, interdependence between renewable-energy and low-carbon assets is also of interest for policymakers, as low-carbon investments could provide adequate incentives to invest in renewable energies and vice versa, thereby determining public funds to be allocated to support the transition to a climate-resilient economy.

We examine dependence between renewable-energy and low-carbon stock prices using a conditional quantile price dependence approach that allows price spillovers between those markets to be explored under different market circumstances, including extreme upward and downward movements in asset values ([1]). Specifically, to assess the impact of price movements of a particular size in one market on stock prices in the other market, we characterize the bivariate dependence structure between renewable-energy and low-carbon stock price returns through copulas, then we compute conditional stock return quantiles and evaluate whether these differ from unconditional quantiles.

The extant literature on renewable-energy and low-carbon stock prices has developed along two separate strands.

One strand has examined the relationship between clean-energy and oil prices. Some studies have explored causality between oil prices and renewables, finding evidence of Granger causality that differs across sample periods and time horizons ([2–6]). Other studies have examined oil price spillovers to renewable stocks, documenting significant impacts from oil price oscillations to renewable stock prices ([7–9]), volatility spillovers between oil and clean-energy stocks ([10–12]) and connectedness between clean energy stocks, oil prices and financial variables ([13]). Likewise, a different set of articles have explored dynamic correlations between renewable energy and stock prices ([14]) and the contribution of energy prices to renewable asset prices and volatility ([15–17]).

The other strand has investigated the effects of carbon emissions on firm performance and on investor portfolios. The authors of [18] find that firm value is negatively impacted by carbon emissions, whereas [19] shows that the cost of capital increases with carbon emissions. The authors of [20,21] find that firms with higher carbon emissions earn higher returns, whereas [22] show that higher emissions are related with higher levels of downside risk. From an investor's perspective, the authors of [23] explores a dynamic investment strategy for passive investors to hedge climate risk without sacrificing financial returns, finding that, even for low-carbon indexes with carbon footprints of 50% less than the benchmark, the tracking error can be virtually eliminated; they also indicate that those results could improve with the pricing of carbon dioxide emissions. Similarly, the authors of [24] shows how bond investor portfolios can be hedged against climate risk with no introduction of unintended exposure that could sacrifice a portfolio's benchmark-tracking properties. More recently, in their investigation of investor portfolio divestment from fossil fuels, the authors of [2] find that clean-energy investments offer better returns, whereas [25], in comparing the financial performance of investment portfolios with and without fossil fuel stocks, report that fossil fuel divestment does not seem to impair portfolio performance, given that fossil fuel stocks do not outperform other stocks on a risk-adjusted basis and that fossil fuel stocks provide relatively limited diversification benefits. Likewise, the authors of [26] contend that socially responsible investing has not been costly in terms of forgone market returns, as the return performance of a fossil-fuel-free portfolio surpasses the S&P 500 returns index due to poor fossil fuel sector performance.

From the investors' perspective, the above-mentioned strands in the literature provide useful information on the impact of energy prices or carbon emissions on the value of low-carbon portfolios composed of either renewable energy or low-carbon assets. However, this literature is silent about the impact of changes in low-carbon asset values on renewable energy asset values and vice versa; such information is crucial for climate-friendly investors as both renewable-energy and low-carbon assets are alternative or complementary assets in terms of the design and risk diversification aims of low-carbon portfolios. This paper fills this gap by analysing interdependence between renewable-energy and low-carbon stock prices in a bivariate copula framework and computing how differently sized stock price movements in one market impact on stock prices in the other market. We model price changes in renewable-energy and low-carbon assets using a multifactor pricing model that includes autoregressive components, with co-movement under different market circumstances modelled through copulas taking into account the effect of common pricing factors in that co-movement. Our empirical study covers the period January 2010 to July 2019 and the European and the USA markets, with renewable-energy stocks represented by the European Renewable Energy and the Wilder Hill

Clean Energy indexes, respectively, and low-carbon assets represented by the Euro STOXX Low Carbon Select 50 and the USA STOXX Low Carbon Select 50 indexes, respectively. Our empirical results point to dissimilarities in both stock markets. Specifically, while we observe interdependence between renewable-energy and low-carbon markets in Europe, those markets do not co-move in the USA. Furthermore, we find evidence of symmetric tail dependence in Europe but independence in the USA. We consistently find evidence of symmetric downside and upside price spillover effects between the European renewable-energy and low-carbon stock markets, differing, however, in that price spillovers from low-carbon to renewable-energy stocks are greater than vice versa. Contrarily, for the USA, we find no evidence of price spillovers.

These findings have implications for both investors and policymakers. Investors holding positions in renewables can hedge such positions using low-carbon assets when they invest in the USA market but should seek alternative hedging devices for the European market. As low-carbon and renewable-energy stocks in the European markets behave as a similar asset class, raising funds for renewables from environmentally-aware investors is more difficult as there are opportunities to invest in other low-carbon assets. Finally, our evidence is informative for the design and funding of renewable energy policies: boosting funding to renewables may have a detrimental effect on low-carbon industries in Europe but only a minor effect in the USA.

The remainder of the paper is laid out as follows. In Section 2 we outline our methodology to assess conditional quantile dependence using copula functions. In Section 3 we describe the main features of our data for renewable-energy and low-carbon stock markets in Europe and the USA. In Section 4 we discuss our results on dependence, the impact of price oscillations from/to low-carbon assets and to/from renewable-energy stocks, and the main implications of those results. Finally, Section 5 summarizes our results and concludes this study.

2. Empirical Methods

2.1. Quantile Dependence Between Renewable-Energy and Low-Carbon Markets

We measure price impacts between the renewable-energy and low-carbon markets using the quantile copula dependence approach developed by [1], which allows the impact of quantile price changes between markets to be assessed. The use of bivariate copula models offers modeling flexibility in featuring bivariate distribution functions, as copulas account for particular data characteristics in the marginal distribution functions, such as time-varying volatilities or leverage effects, and they allow dependence to differ under different market circumstances, in particular in times of extreme price oscillations.

To begin with, using copulas rather than quantile regression results in greater modeling flexibility; this is because copulas enable heterogeneity in characterizing marginal distributions and also account for specific data features such as conditional heteroskedasticity, volatility asymmetries, and leverage effects. Moreover, our empirical setup allows for time-varying dependence, so the impact of oil price changes on stock returns are allowed to differ in different moments of time depending on the dependence and volatility features of the corresponding markets.

Let re_t and lc_t be the (log) change in prices of renewable-energy and low-carbon stocks, respectively. The impact of a change in the price of a low-carbon asset of a size given by its β-quantile on the α-quantile of the renewable-energy market can be measured by the conditional -quantile of the renewable return distribution at time t, $q_{\alpha,\beta,t}^{re_t|lc_t}$, as:

$$P\left(r_t \leq q_{\alpha,\beta,t}^{re_t|lc_t} | lc_t \leq q_{\beta,t}^{lc_t}\right) = \alpha, \tag{1}$$

where $q_{\beta,t}^{lc_t}$ is the unconditional β-quantile of the low-carbon price returns distribution: $P\left(lc_t \leq q_{\beta,t}^{lc_t}\right) = \beta$. From this conditional quantile, we can quantify how price fluctuations in low-carbon stocks of different sizes impact on renewable-energy stocks under different market scenarios as reflected by the quantiles of low-carbon stocks. Similarly, we can obtain the reverse impact, i.e., the impact of price fluctuations in renewable-energy stock prices on the prices of low-carbon assets.

From [27]'s theorem on equality between the joint distribution function and a copula function C, we can express Equation (1) in terms of the joint distribution function or the copula function as:

$$F_{re_t lc_t}\left(q_{\alpha,\beta,t}^{re_t|lc_t}, q_{\beta,t}^{lc_t}\right) = C\left(F_{re_t}\left(q_{\alpha,\beta,t}^{re_t|lc_t}\right), F_{lc_t}\left(q_{\beta,t}^{lc_t}\right)\right) = \alpha\beta, \tag{2}$$

where $F_{re_t}(\cdot)$ and $F_{lc_t}(\cdot)$ denote the marginal distribution functions for the renewable-energy and low-carbon price changes, respectively and where the second equality follows from the fact that the joint distribution is the product of the conditional and marginal distributions, $F_{re_t lc_t}\left(q_{\alpha,\beta,t}^{re_t|lc_t}, q_{\beta,t}^{lc_t}\right)$ $F_{re_t}\left(q_{\beta,t}^{lc_t}\right)$, with $F_{re_t lc_t}\left(q_{\alpha,\beta,t}^{re_t|lc_t}, q_{\beta,t}^{lc_t}\right) = \alpha$ and $F_{re_t}\left(q_{\beta,t}^{lc_t}\right) = \beta$. Hence, for given values for α and β and for the copula model specification, we can compute $q_{\alpha,\beta,t}^{re_t|lc_t}$ by inverting the copula function in Equation (2), $C\left(F_{re_t}\left(q_{\alpha,\beta,t}^{re_t|lc_t}\right), \beta\right) = \alpha\beta$, in order to obtain the value of $\hat{F}_{re_t}\left(q_{\alpha,\beta,t}^{re_t|lc_t}\right)$; then, by inverting the marginal distribution function of re_t we obtain the conditional quantile as:

$$q_{\alpha,\beta,t}^{re_t|lc_t} = F_{re_t}^{-1}\left(\hat{F}_{re_t}\left(q_{\alpha,\beta,t}^{re|lc_t}\right)\right). \tag{3}$$

Note that if renewable-energy and low-carbon stock markets are independent, then $C\left(F_{re_t}\left(q_{\alpha,\beta,t}^{re|lc_t}\right), \beta\right) = F_{re_t}\left(q_{\alpha,\beta,t}^{re_t|lc_t}\right)\beta$, so $q_{\alpha,\beta,t}^{re_t|lc_t} = q_{\alpha,t}^{re_t}$. Hence, the difference between conditional and unconditional renewable-energy return quantiles provides information on the impact of low-carbon stock price changes on renewable-energy stock returns.

To compute the conditional quantile through copulas we need information on the marginal distribution models and on dependence between renewable-energy and low-carbon market prices as given by the copula function. Using copulas rather than the conditional marginal distribution to compute conditional quantiles has the appeal of flexibility, in that copulas separate modeling of the marginals and of the dependence structures, and they capture dependence in the case of sharp upward (upper quantiles) or downward (lower quantiles) price movements.

2.2. Marginal and Copula Models

As the mean and variance of financial return series exhibit time-varying behavior and stock returns depend on general pricing factors, we estimate the price dynamics of renewable-energy and low-carbon stocks using an autoregressive moving average (ARMA) model with p and q lags and with exogenous variables as given by the five pricing factors proposed by [28,29]:

$$y_t = \phi_0 + \sum_{j=1}^{p} \phi_j y_{t-j} + \sum_{h=1}^{q} \varphi_j \varepsilon_{t-h} + \beta_1 MKT_t + \beta_2 SMB_t + \beta_3 HML_t + \beta_4 RMW_t + \beta_5 CMA_t + \varepsilon_t, \tag{4}$$

where y_t denotes the excess price returns in renewable-energy and low-carbon stocks and where the pricing factors are as follows: MKT_t is the excess return of the market portfolio; SMB_t is the difference between the returns of a diversified portfolio comprised of small and large assets; HML_t is the difference between high book-to-market and low book-to-market portfolio returns; RMW_t is the difference between returns for a diversified portfolio of robust and weak profitability assets; and CMA_t is the difference between portfolio returns for low (conservative) and high (aggressive) investment firms. ε_t is a stochastic component with zero mean and variance σ_t^2, which has a dynamic described by a threshold generalized autoregressive conditional heteroskedasticity (TGARCH) model:

$$\sigma_t^2 = \omega + \sum_{k=1}^{r} \theta_k \sigma_{t-k}^2 + \sum_{h=1}^{m} \alpha_h \varepsilon_{t-h}^2 + \sum_{h=1}^{m} \lambda_h 1_{t-h} \varepsilon_{t-h}^2, \tag{5}$$

where ω is a constant parameter and where the parameters θ and α account for the generalized autoregressive conditional heteroskedasticity (GARCH) and autoregressive conditional heteroskedasticity (ARCH) effects, respectively. $1_{t-h} = 1$ for $\varepsilon_{t-h} < 0$, then the parameter λ captures asymmetric effects: when $\lambda > 0$ ($\lambda < 0$) negative shocks have more (less) impact on variance than positive shocks (note that for $\lambda = 0$ we have symmetric effects as given by the GARCH model). Fat tails and asymmetries of the stochastic component ε_t, and thus of y_t, are captured by [30] skewed-t density distribution; this distribution is characterized by parameters v (the degrees-of-freedom parameter, $2 < v < \infty$) and η (the symmetric parameter, $-1 < \eta < 1$).

We model dependence by considering different copula specifications for the variables x and y, with $u = F_x(x)$ and $v = F_y(y)$. Specifically, we capture positive and negative dependence using the bivariate Gaussian copula, given by $C_N(u, v; \rho) = \Phi\left(\Phi^{-1}(u), \Phi^{-1}(v)\right)$, where Φ is the bivariate standard normal cumulative distribution function with correlation ρ and where $\Phi^{-1}(u)$ and $\Phi^{-1}(v)$ are standard normal quantile functions. Similarly, positive and negative dependence is captured by the student-t copula, which is given by $C_{ST}(u, v; \rho, v) = T\left(t_v^{-1}(u), t_v^{-1}(v)\right)$, where T is the bivariate student-t cumulative distribution function with the degree-of-freedom parameter v and dependence given by the correlation coefficient ρ and where $t_v^{-1}(u)$ and $t_v^{-1}(v)$ are the quantile functions of the univariate student-t distribution. Gaussian and student–t copulas differ in terms of their tail dependence: the former exhibit zero tail dependence while the latter show symmetric tail dependence and converge to the Gaussian when the degrees of freedom go to infinity. We also consider the Gaussian and student-t copulas with time-varying parameters, with a dynamic given by [31]:

$$\rho_t = \Lambda\left(\psi_0 + \psi_1\rho_{t-1} + \psi_2\frac{1}{q}\sum_{j=1}^{q}\Phi^{-1}\left(u_{t-j}\right)\cdot\Phi^{-1}\left(v_{t-j}\right)\right),\tag{6}$$

where $\Lambda(x) = (1 - e^{-x})(1 + e^{-x})^{-1}$ is the modified logistic transformation that retains ρ_t in $(-1,1)$. As for the student-t copula, $\Phi^{-1}(x)$ is replaced by $t_v^{-1}(x)$. We also use the symmetric Plackett copula, which, like the Gaussian copula, exhibits tail independence although it displays more dependence for large joint realizations. It is given by:

$$C_P(u, v; \theta) = \frac{1}{2(\theta - 1)}(1 + (\theta - 1)(u + v)) - \sqrt{(1 + (\theta - 1)(u + v))^2 - 4\theta(\theta - 1)uv}.\tag{7}$$

Furthermore, we capture asymmetric dependence using the Gumbel copula, given by $C_G(u, v; \delta) = \exp\left(-\left((-\log u)^\delta + (-\log v)^\delta\right)^{1/\delta}\right)$, which has upper tail dependence and lower tail independence. Moreover, we rotate the Gumbel copula 180° with parameter $\delta > 0$: $C_{RG180}(u, v; \delta) = v - \exp\left(-\left((-\log(1 - u))^\delta + (-\log v)^\delta\right)^{1/\delta}\right)$. Finally, we also consider time-varying dynamics of the dependence parameter as given by:

$$\delta_t = \omega + \beta\delta_{t-1} + \alpha\frac{1}{q}\sum_{j=1}^{q}\left|u_{t-j} - v_{t-j}\right|\tag{8}$$

Finally, the parameters of the marginal and copula models are estimated using the inference function for margins ([32]), which allows parameter estimation in two steps. First, the parameters of the marginal models are first estimated using maximum likelihood. Next, the copula parameters are estimated by maximum likelihood using, as pseudo-sample observations for the copulas, the probability integral transformation of the standardized residuals from the marginals. The number of lags in the mean and variance equations in the marginal models are selected using the Akaike information criteria (AIC), whereas the adequate copula specification is selected using the AIC adjusted for small-sample bias ([33]).

3. Data

Our empirical analysis on quantile dependence between renewable-energy and low-carbon stocks in the European and USA stock markets is based on the respective stock market indices. Specifically, we consider the Europe STOXX Low Carbon Select 50 Index (LC-EU) and the USA STOXX Low Carbon Select 50 Index (LC-USA) for the European and USA stock markets, respectively. These indices, which exclude all companies involved in the coal sector, capture the performance of low carbon emissions stocks with low volatility and high dividends selected from the universe of companies included in the STOXX Europe 600 Index and in the STOXX Global 1800 index, respectively. The 50 assets included in the index are weighted according to the inverse of their volatility, with a cap at 10%, and the index is reviewed quarterly.

To account for the performance of renewable-energy stocks, we take the European Renewable Energy index (ERIX) and the Wilder Hill Clean Energy index (ECO) for Europe and USA, respectively. ERIX is comprised of the largest European renewable-energy companies with wind, solar, biomass, and water energy generation as their main activities. ECO is an equal-dollar-weighted index of a set of companies that develop activities related to clean energies and conservation.

The sample covered the period 1 January 2010 to 31 July 2019, with the start date of the sample period determined by the availability of data for low-carbon indices. Data was sourced from Bloomberg on a daily basis. Figure 1 displays the temporal dynamics of renewable-energy and low-carbon markets in the European and USA markets, showing that these markets follow similar trends, with price changes harmonized in Europe, but not synchronized in the USA. We computed daily price returns as the first difference for the (log) value of those indices. Table 1 presents the main statistical features of the daily returns. For all markets under study, average daily returns are close to zero and standard deviations are greater for the renewable-energy stocks than for the low-carbon stocks. Higher volatility in renewable-energy markets is also confirmed by the maximum and minimum values of returns. All price returns exhibit negative skewness and the price return distributions have fat tails. In fact, the Jarque–Bera (JB) test rejects normality. The evidence of serial dependence provided by the Ljung–Box (LJ) statistic is mixed: although most of the series exhibit serial independence, low-carbon series for Europe show serial dependence. Finally, the ARCH test points to the presence of conditional heteroskedasticity in the series.

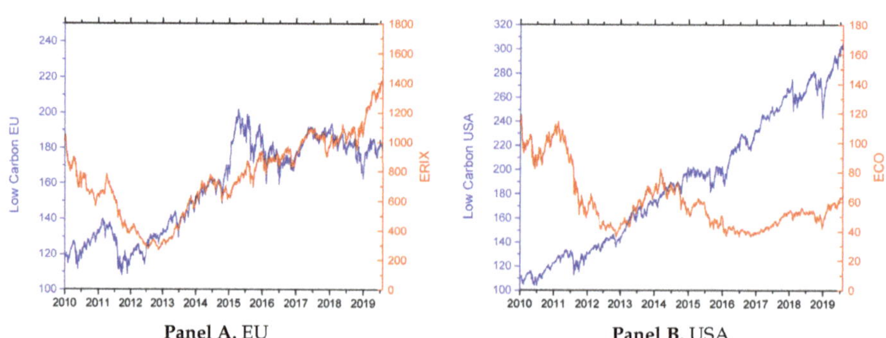

Figure 1. Time series plot for daily renewable-energy and low-carbon indices.

Data for the pricing factors and the risk-free interest rates to compute excess returns in renewable-energy and low-carbon markets in Europe and the USA were sourced from the Kenneth French data library (http://mba.tuck.dartmouth.edu/pages/faculty/ken.french/data_library.html). Table 2 presents descriptive statistics for those pricing factors in both markets, showing that their behaviors differ across stock markets.

Table 1. Descriptive statistics for renewable-energy and low-carbon indices.

	LC-EU	ERIX	LC-USA	ECO
Mean	0.015	0.003	0.042	−0.024
Maximum	4.966	6.622	4.454	8.239
Minimum	−5.706	−7.978	−6.144	−10.21
Std. Dev.	0.823	1.492	0.773	1.673
Skewness	−0.338	−0.301	−0.435	−0.27
Kurtosis	6.887	5.162	8.185	5.214
JB	1563.09 *	505.59 *	2775.70 *	521.49 *
ARCH	19.137	8.336	29.129	22.173
	[0.000]	[0.000]	[0.000]	[0.000]
Q(20)	45.322	28.632	33.629	41.855
	[0.001]	[0.095]	[0.029]	[0.003]

Note. The table presents descriptive statistics for price returns in low-carbon markets in the EU (LC-EU) and the USA (LC-USA) and in the corresponding renewable-energy markets, ERIX and ECO, respectively. Data cover daily periods from 1 January 2010 to 31 July 2019. JB denotes the Jarque–Bera statistic for the null hypothesis of normality; an asterisk denotes rejection of the null hypothesis. Finally, ARCH denotes Engle's Lagrange multiplier test for conditional heteroskedasticity and Q(20) denotes Ljung–Box statistics for serial correlation in the price return series. Both tests were computed with 20 lags and their p values are reported in square brackets.

Table 2. Descriptive statistics for pricing factors.

Panel A. EU				
	MKT	SMB	HML	RMW	CMA

	MKT	SMB	HML	RMW	CMA
Mean	0.023	0.004	−0.011	0.017	0.000
Maximum	6.850	3.210	3.760	1.630	1.150
Minimum	−8.800	−2.250	−2.130	−1.910	−1.070
Std. Dev.	1.098	0.443	0.443	0.296	0.245
Skewness	−0.323	−0.084	0.329	−0.372	0.130
Kurtosis	7.741	5.949	6.407	5.299	4.014
JB	2298.54 *	876.03 *	1208.94 *	586.37 *	109.94 *
ARCH	16.393	11.621	10.14	3.428	8.021
	[0.000]	[0.000]	[0.000]	[0.000]	[0.000]
Q(20)	41.574	41.41	33.372	37.808	33.017
	[0.003]	[0.003]	[0.031]	[0.009]	[0.034]

Panel B. US				

	MKT	SMB	HML	RMW	CMA
Mean	0.052	−0.001	−0.01	0.006	0.000
Maximum	5.060	3.620	2.390	1.660	1.950
Minimum	−6.970	−1.990	−1.83	−1.63	−1.320
Std. Dev.	0.963	0.518	0.497	0.342	0.303
Skewness	−0.402	0.177	0.303	0.034	0.361
Kurtosis	7.410	4.639	4.494	4.567	4.952
JB	2018.24 *	282.48 *	261.03 *	246.89 *	434.88 *
ARCH	28.873	7.701	10.781	12.382	7.396
	[0.000]	[0.000]	[0.000]	[0.000]	[0.000]
Q(20)	50.511	22.075	21.486	21.031	19.639
	[0.000]	[0.336]	[0.369]	[0.395]	[0.481]

Note. The table presents descriptive statistics for pricing factors, MKT, SMB, HML, RMW and CMA, in the EU and USA markets for the period 1 January 2010 to 31 July 2019. JB denotes the Jarque–Bera statistic for the null hypothesis of normality; rejection of the null hypothesis is indicated with an asterisk. Finally, ARCH denotes Engle's Lagrange multiplier test for conditional heteroskedasticity and Q(20) denotes Ljung–Box statistics for serial correlation in the price return series. Both tests were computed with 20 lags and their p values are reported in square brackets.

4. Empirical Evidence

4.1. Results for Marginal and Copula Models

Parameter estimates and goodness-of-fit tests for the marginal models for renewable-energy and low-carbon indices in the EU and the USA are presented in Table 3. We selected suitable lags for the mean and variance by considering lag values between 0 and 2, taking as the optimal values those that minimized the AIC. Our estimates reflect serial dependence in all price return series, given the significant autoregressive and moving average coefficients. Parameter estimates for the pricing factors indicate that all return series are dependent on the market factor, with betas below one indicating that renewable-energy and low-carbon stocks are defensive stocks, with the exception of the ECO index. However, we find mixed evidence for the remaining pricing factors as the significance of those factors differs across markets. Likewise, parameter estimates for the volatility dynamics indicate that volatility displays persistence and no leverage effects, with the exception of the European low-carbon market. The degrees-of-freedom parameter also indicates that the error terms are generally symmetric and exhibit fat tails, whereas asymmetry is significant in the low-carbon markets.

The last six columns of Table 3 show results for goodness-of-fit tests for the estimated marginal models. The LJ test indicates that there is no serial correlation in either the residual series or the squared residual series, and the ARCH-Lagrange multiplier (ARCH-LM) statistic indicates that no GARCH effects remain in the model residuals. In comparing the empirical and theoretical distribution functions of the standardized residuals, the Kolmogorov–Smirnov (KS), Cramér–von Mises (CVM), and Anderson–Darling (AD) tests all support the null hypothesis of correct specification of the distribution models for all the series.

We estimate copula model parameters using the probability integral transform of the standardized residuals from the estimated marginal models as pseudo-sample observations for the copula. Parameter estimates for the static and time-varying copulas are reported in Table 4. Empirical estimates point to relevant difference between the European and USA markets. Thus, while in the European market we find evidence of positive dependence between renewable-energy and low-carbon stock markets, for the USA we find that this dependence to be negative and small. Evidence on comparing copulas through the AIC values indicates that the static student-t copula provides the best fit for the European markets and the Plackett copula for the USA market. Furthermore, dependence between renewable-energy and low-carbon stock markets is fundamentally static. We only find evidence of tail dependence in the European market, so upward or downward movements in renewable-energy stock prices have impacts on the low-carbon market and vice versa. In contrast, for the USA market we find evidence of no tail dependence and weak negative average dependence, so abrupt price changes in renewable-energy stock prices have negligible effects on low-carbon assets and vice versa.

4.2. Price Impact Results for the Renewable-Energy and Low-Carbon Stock Markets

We estimate conditional quantiles using information from the estimated marginal and copula models, taking different values for the quantiles α and β given by 0.05, 0.10, 0.25, 0.5, 0.75, 0.9, and 0.95. To assess the relative impact of low-carbon stock prices on renewable-energy prices and vice versa, we also estimate the unconditional quantiles from the marginal models as $q_{\beta,t}^{y_t} = \mu_t + F_{v,\eta}^{-1}(\alpha)\sigma_t$, for $y_t = re_t, lc_t$, and μ_t and σ_t are given by the ARMA and GARCH components of the marginal model, with $F_{v,\eta}^{-1}(\alpha)$ denoting the value of the α-quantile of the skewed student-t distribution.

Figure 2 depicts the quantile dynamics of the upper and lower conditional and unconditional renewable quantiles in the European and the USA markets, considering the impact of high (low) price fluctuations in low-carbon stocks as given by the 0.9 (0.1) quantile on the high (low) renewable-energy quantile as given by the 0.9 (0.1) quantile. Consistent with the evidence on tail dependence in the European renewable-energy stock market, we found that differences between conditional and unconditional quantiles in the upper and lower tails of the joint distribution were sizable and of a similar size. Hence, sharp upward or downward movements in low-carbon stocks have an impact on prices of renewable-energy stocks in the European markets. However, this effect is not observed in the USA market, as there is near zero dependence, i.e., the impact of price oscillations in low-carbon assets has no sizeable impact on renewable-energy stock prices, as reflected in Panel B of Figure 2.

As for the impact of price oscillations in renewable-energy stocks on low-carbon stock prices, Figure 3 depicts upper and lower conditional and unconditional low-carbon quantiles in the European and the USA markets, considering the impact of high (low) price fluctuations in renewable-energy stocks as given by the 0.9 (0.1) quantile on the high (low) low-carbon quantile as given by the 0.9 (0.1) quantile. Graphical evidence reflects that price impacts differ in both stock markets; in the European market, price movements in renewable-energy stocks have a significant impact on low-carbon stock prices, whose impact is smaller than in reverse, whereas in the USA market—consistently with near independence—differences between conditional and unconditional quantiles are small.

Table 3. Maximum likelihood estimates.

	LC-EU	ERIX	LC-USA	ECO
Mean				
ϕ_0	0.001	−0.011	−0.001	−0.072 *
	(0.119)	(−0.499)	(−0.208)	(−3.807)
ϕ_1	0.358 *	0.000	0.627 *	0.046 *
	(2.353)	(0.004)	(5.616)	(2.171)
ϕ_2		0.048 *		0.042 *
		(2.453)		(1.964)
φ_0	−0.320 *		−0.622 *	
	(−2.083)		(−6.010)	
MKT	0.447 ⁺	0.799 *	0.805 *	1.150 ⁺
	(21.150)	(24.330)	(85.140)	(49.020)
SMB	−0.553 *	−0.115	−0.163 *	0.782 *
	(−14.070)	(−1.540)	(−12.090)	(20.130)
HML	−0.186 *	0.186	0.014	0.081
	(−4.248)	(1.770)	(0.800)	(1.802)
RNW	0.036	0.005	0.250 *	−0.433 *
	(0.651)	(0.036)	(11.520)	(−6.985)
CMA	0.056	−0.269 *	0.350 *	−0.243 *
	(1.086)	(−2.411)	(12.270)	(−3.364)
Variance				
ω	0.016 *	0.133	0.029 *	0.019
	(2.958)	(1.613)	(2.333)	(1.677)
α_1	0.055 *	(0.064 *	0.181 *	0.043 *
	(2.897)	(2.761)	(3.428)	(2.629)
β_1	0.831 *	0.253	0.293 *	0.936 *
	(19.820)	(1.117)	(2.394)	(34.420)
β_2		0.562 *	0.223	
		(2.234)	(1.858)	
λ	0.068 *	0.027	−0.049	−0.007
	(1.966)	(0.673)	(−0.770)	(−0.620)
Asymetry	−0.111 *	0.028	−0.067 *	0.051
	(−3.698)	(0.935)	(−2.100)	(1.720)
Tail	9.314 *	6.296 *	7.140 *	7.728 *
	(5.774)	(7.729)	(7.138)	(6.532)
LogLik	−1397.5	−3562.74	−396.486	−3039.77
LJ	16.590	29.485	23.019	20.784
	[0.55]	[0.05]	[0.19]	[0.29]
LJ(2)	10.464	14.456	21.975	16.065
	[0.92]	[0.63]	[0.19]	[0.59]
ARCH	0.522	0.746	1.118	0.749
	[0.96]	[0.78]	[0.32]	[0.78]
KS	[0.90]	[0.96]	[0.86]	[0.98]
CVM	[0.99]	[0.99]	[0.88]	[0.99]
AD	[0.99]	[0.99]	[0.91]	[0.99]

Note. The table presents parameter estimates and z-statistics (in brackets) for the marginal models described in Equations (4) and (5). An asterisk (*) indicates significance at 5%. LogLik denotes the log-likelihood value. LJ, LJ(2) and ARCH, respectively, denote the Ljung–Box statistic for serial correlation in the residual model and the squared residual model and Engle's Lagrange multiplier test for the ARCH effect in residuals computed with 20 lags. KS, CVM, and AD, respectively, denote the Kolmogorov–Smirnov, Cramér–von Mises and Anderson–Darling statistics for the null hypothesis of correct model specification. Rejection of the null hypothesis is indicated with *p* values (in square brackets) below 0.05.

Table 4. Estimates for the copula models.

Panel A. Parameter Estimates for Time-Invariant Copulas.		
	EU	**USA**
Gaussian copula		
ρ	0.377 *	−0.075 *
	(0.01)	(0.02)
AIC	−367.773	−11.515
Student−t copula		
ρ	0.393 *	−0.078 *
	(0.02)	(0.02)
v	7.228 *	43.210
	(0.54)	(49.14)
AIC	**−415.622**	−11.287
Gumbel copula		
δ	1.305 *	1.000 *
	(0.02)	(0.02)
AIC	−344.293	2.002
Rotate Gumbel copula		
δ	1.315 *	1.001 *
	(0.02)	(0.00)
AIC	−381.959	1.842
Plackett copula		
θ	3.461 *	0.780 *
	(0.20)	(0.05)
AIC	−396.806	**−14.940**

Panel B. Parameter Estimates for Time-Invariant Copulas.		
	EU	**USA**
TVP-Gaussian		
ψ_0	−0.029	−0.235 *
	(0.02)	(0.07)
ψ_1	0.033 *	0.191
	(0.02)	(0.13)
ψ_2	2.155 *	−1.340 *
	(0.07)	(0.40)
AIC	−378.553	−9.807
TVP−Student		
ψ_0	1.777 *	−0.266 *
	(0.10)	(0.09)
ψ_1	−0.010	0.061
	(0.01)	(0.08)
ψ_2	−2.362 *	−1.370 *
	(0.04)	(0.64)
v	7.264 *	20.000 *
	(1.13)	(4.81)
AIC	−412.78	−5.978
TVP−Gumbel		
$\bar{\omega}$	0.932	0.000
	(0.49)	(1.00)
$\bar{\beta}$	−0.212	0.000
	(0.35)	(1.00)
$\bar{\alpha}$	−0.408 *	0.000
	(0.2)	(1.00)
AIC	−346.097	6.042

Table 4. *Cont.*

Panel B. Parameter Estimates for Time-Invariant Copulas.		
	EU	**USA**
TVP−Rotate gumbel		
$\bar{\omega}$	1.031 *	5.000
	(0.39)	(473.32)
$\bar{\beta}$	−0.281	−4.948
	(0.31)	(470.40)
$\bar{\alpha}$	−0.398 *	−0.045
	(0.14)	(5.61)
AIC	−383.263	5.831

Note. The table presents parameter estimates for different copula models along with their standard errors reported in brackets. An asterisk (*) indicates significance of the parameter at 5%. The best copula fit is given by the copula model that attains the minimum Akaike information criterion (AIC) value adjusted for small-sample bias, indicated in bold. For the time-varying parameter (TVP) copulas, the value of q was set to 10.

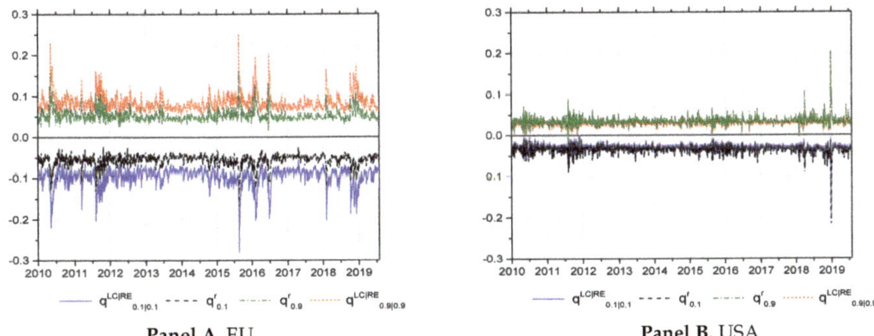

Figure 2. Temporal dynamics for upper and lower conditional and unconditional quantiles of renewable-energy stock returns.

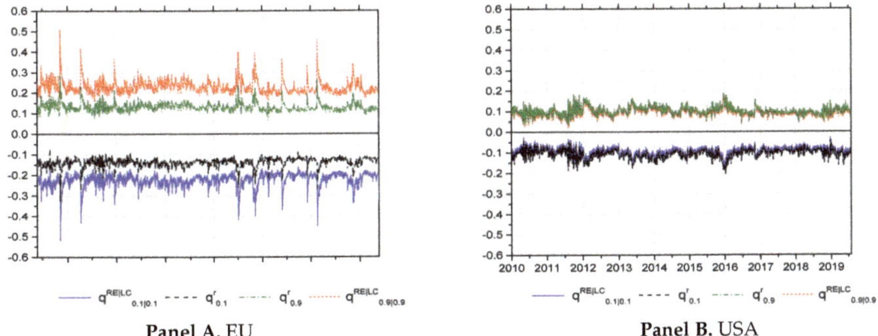

Figure 3. Temporal dynamics for upper and lower conditional and unconditional quantiles of low-carbon stock returns.

Finally, Figures 4 and 5 summarize the relative impact of price changes in low-carbon assets of specific sizes on renewable-energy stocks and vice versa, respectively. For different values, the plots represent the average value of the conditional quantile over the unconditional quantile: values greater than one, depicted in warm colours, indicate that stock price changes in one market affect the corresponding unconditional quantile of the other market, whereas values in cold colours indicate the opposite. For the European markets, Panel A in Figures 4 and 5 confirms that renewables and low carbon markets closely co-move, so upward or downward movements in one of the markets have a positive and significant effect on the prices in the other market. Likewise, graphical evidence also corroborates that opposite movements in renewable-energy and low-carbon prices are not related, consistent with the idea that markets move in tandem. In contrast, for the US, graphical evidence in Panel B reflects the fact that renewable-energy and low-carbon markets move independently, i.e., price changes in one market are not reflected in price movements in the other market as indicated by the equality between conditional and unconditional quantiles.

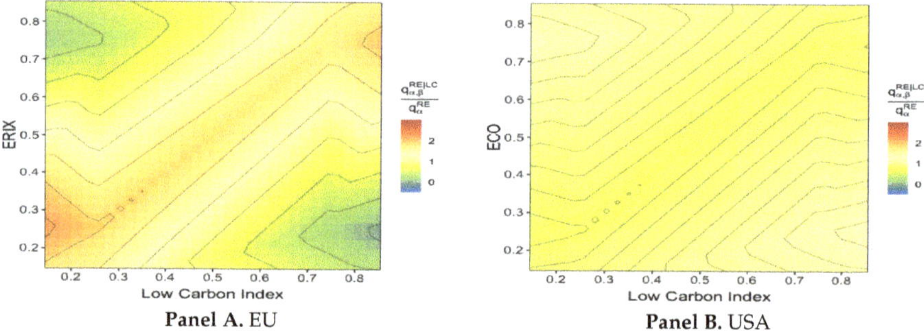

Panel A. EU   **Panel B.** USA

Figure 4. Average value of conditional over unconditional quantiles for renewable-energy stocks.

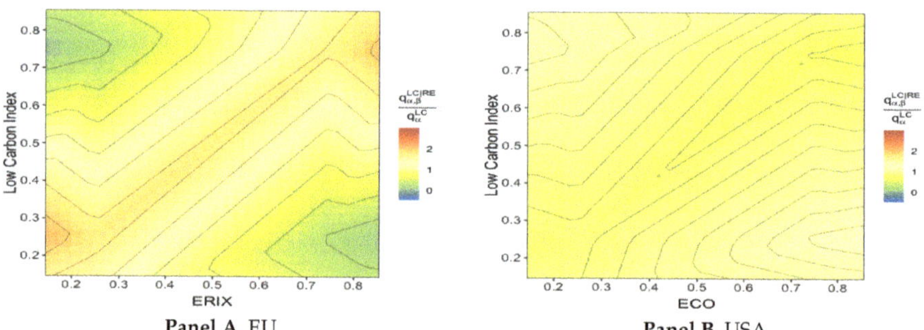

Panel A. EU   **Panel B.** USA

Figure 5. Average value of conditional over unconditional quantiles for low-carbon stocks.

5. Conclusions

Investing in renewable-energy and low-carbon assets is a straightforward way for investors to align their portfolios with a low-carbon and more climate-resilient economy. However, the diversification benefits of holding positions in such firms closely depends on the way both kinds of assets co-move. Climate-friendly investors could take advantage of this information to build more adequate portfolio investment strategies. We therefore explored interdependence between prices of renewable-energy and low-carbon assets in the European and USA stock markets for the period 2010–2019. We use copula functions, which report information on dependence under different market circumstances, even though we have no information on the causality effects.

Our empirical evidence documents that European, but not USA, renewable-energy and low-carbon markets co-move and, likewise, we find evidence of symmetric tail dependence in Europe but tail independence in the USA. Consistently, we find that upside or downside movements in the prices of low-carbon assets impact on renewable-energy asset prices in Europe, while the reverse is also notable, although smaller in size. In contrast, for the USA market we find that the impact of upward or downward price oscillations in low-carbon asset prices on the price of renewable-energy stocks is not sizable as is the case with the reverse.

Our empirical findings have practical implications for investor decision-making and for policy-makers as follows. First, evidence of positive co-movement between renewable-energy and low-carbon asset prices in Europe indicates that, as both kind of assets move in the same direction under different market scenarios, long positions in low-carbon assets cannot be hedged using long positions in renewable-energy assets, and vice versa. In the USA market, in contrast, since renewable-energy and low-carbon stock prices move independently, investors in either market could use one set of assets to hedge financial positions in the other market. Second, in Europe, climate-friendly investors cannot use renewable-energy assets to manage downside risks for long positions in low-carbon assets, and vice versa, unlike investors in the USA. Third, our evidence on interdependence between renewable-energy and low-carbon asset prices is useful for the design of energy policies to support and fund renewable energy investments. Specifically, when renewable-energy and low-carbon asset prices co-move—as happens in the European markets—public funding of renewables impact on renewable-energy companies, and this impact leads to price externalities for low-carbon companies. Likewise, the withdrawal of support policies for renewable energies (such as subsides) will have negative effects on the price of renewable-energy stocks that will be transmitted to the price of low-carbon assets. As a result, policy decisions regarding energy transition to a decarbonized economy should take into account the effects on low-carbon companies, as also crucial in the transition to a climate-resilient economy. However, when low-carbon and renewable-energy markets move independently—as happens in the USA—such policy effects are irrelevant. Finally, in raising funds for the transition to a low-carbon economy, the dependence between climate-friendly assets such as renewable-energy and low-carbon stocks is such as to render them a similar asset class; therefore, funds for renewable energies face competition in the demand for funds for other low-carbon industries that may also be attractive to environmentally friendly investors—the case in Europe; in the USA, however, the independence of the asset classes may spur investment incentives in climate-friendly assets such as renewable-energy and low-carbon stocks.

Author Contributions: Conceptualization, J.C.R., A.U., Y.C.; Methodology, J.C.R., A.U.; Data collection, A.U., Y.C., Estimation, A.U., Y.C.; Writing—Original Draft Preparation, J.C.R.; Review & Editing, J.C.R., A.U., Y.C.; Funding Acquisition, J.C.R., A.U.

Funding: Financial support from the Agencia Estatal de Investigación (Ministerio de Ciencia, Innovación y Universidades) is acknowledged for research project RTI2018-100702-B-I00, co-funded by the European Regional Development Fund (ERDF/FEDER). Juan C. Reboredo acknowledges funding from the Xunta de Galicia for research project CONSOLIDACIÓN 2019 GRC GI-2060 Análise Económica dos Mercados e Institucións—AEMI (ED431C 2019/11). Andrea Ugolini acknowledges financial support from the Brazilian National Council for Scientific and Technological Development (CNPq).

Conflicts of Interest: The authors declare no conflict of interest.

References

1. Reboredo, J.C.; Ugolini, A. Quantile dependence of oil price movements and stock returns. *Energy Econ.* **2016**, *54*, 33–49. [CrossRef]
2. Henriques, I.; Sadorsky, P. Investor implications of divesting from fossil fuels. *Glob. Financ. J.* **2018**, *38*, 30–44. [CrossRef]
3. Managi, S.; Okimoto, T. Does the price of oil interact with clean energy prices in the stock market? *Jpn. World Econ.* **2013**, *27*, 1–9. [CrossRef]
4. Kumar, S.; Managi, S.; Matsuda, A. Stock prices of clean energy firms, oil and carbon markets: A vector autoregressive analysis. *Energy Econ.* **2012**, *34*, 215–226. [CrossRef]
5. Broadstock, D.C.; Cao, H.; Zhang, D. Oil shocks and their impact on energy related stocks in China. *Energy Econ.* **2012**, *34*, 1888–1895. [CrossRef]
6. Reboredo, J.C.; Rivera-Castro, M.A.; Ugolini, A. Wavelet-based test of co-movement and causality between oil and renewable energy stock prices. *Energy Econ.* **2017**, *61*, 241–252. [CrossRef]

7. Inchauspe, J.; Ripple, R.D.; Trück, S. The dynamics of returns on renewable energy companies: A state-space approach. *Energy Econ.* **2015**, *48*, 325–335. [CrossRef]

8. Reboredo, J.C. Is there dependence and systemic risk between oil and renewable energy stock prices? *Energy Econ.* **2015**, *48*, 32–45. [CrossRef]

9. Bondia, R.; Ghosh, S.; Kanjilal, K. International crude oil prices and the stock prices of clean energy and technology companies: Evidence from non-linear cointegration tests with unknown structural breaks. *Energy* **2016**, *101*, 558–565. [CrossRef]

10. Sadorsky, P. Correlations and volatility spillovers between oil prices and the stock prices of clean energy and technology companies. *Energy Econ.* **2012**, *34*, 248–255. [CrossRef]

11. Sadorsky, P. Modeling renewable energy company risk. *Energy Policy* **2012**, *40*, 39–48. [CrossRef]

12. Wen, X.; Guo, Y.; Wei, Y.; Huang, D. How do the stock prices of new energy and fossil fuel companies correlate? Evidence from China. *Energy Econ.* **2014**, *41*, 63–75. [CrossRef]

13. Ferrer, R.; Jawad, S.; Shahzad, H.; López, R. Time and frequency dynamics of connectedness between renewable energy stocks and crude oil prices. *Energy Econ.* **2018**, *76*, 1–20. [CrossRef]

14. Zhang, G.; Du, Z. Co-movements among the stock prices of new energy, high-technology and fossil fuel companies in China. *Energy* **2017**, *135*, 249–256. [CrossRef]

15. Dutta, A. Oil price uncertainty and clean energy stock returns: New evidence from crude oil volatility index. *J. Clean. Prod.* **2019**, *164*, 1157–1166. [CrossRef]

16. Reboredo, J.C.; Ugolini, A. The impact of energy prices on clean energy stock prices. A multivariate quantile dependence approach. *Energy Econ.* **2018**, *76*, 136–152. [CrossRef]

17. Sun, C.; Ding, D.; Fang, X.; Zhang, H.; Li, J. How do fossil energy prices affect the stock prices of new energy companies? Evidence from Divisia energy price index in China's market. *Energy* **2019**, *135*, 2637–645. [CrossRef]

18. Matsumura, E.M.; Prakash, R.; Vera-Muñoz, S.C. Firm-value effects of carbon emissions and carbon disclosures. *Account. Rev.* **2013**, *89*, 695–724. [CrossRef]

19. Chava, S. Environmental externalities and cost of capital. *Manag. Sci.* **2014**, *60*, 2223–2247. [CrossRef]

20. Bolton, P.; Kacperczyk, M.T. Do Investors Care about Carbon Risk? Available online: https://ssrn.com/abstract=3398441 (accessed on 16 October 2019).

21. Hsu, P.H.; Li, K.; yang Tsou, C. The Pollution Premium. *J. Financ. Econ.* forthcoming.

22. Ilhan, E.; Sautner, Z.; Vilkov, G. Carbon Tail Risk. Available online: https://ssrn.com/abstract=3204420 (accessed on 12 September 2019).

23. Andersson, M.; Bolton, P.; Samama, F. Hedging climate risk. *Financ. Anal. J.* **2016**, *72*, 13–32. [CrossRef]

24. de Jong, M.; Nguyen, A. Weathered for climate risk: A bond investment proposition. *Financ. Anal. J.* **2016**, *72*, 34–39. [CrossRef]

25. Trinks, A.; Scholtens, B.; Mulder, M.; Dam, L. Fossil fuel divestment and portfolio performance. *Ecol. Econ.* **2018**, *146*, 740–748. [CrossRef]

26. Halcoussis, D.; Lowenberg, A.D. The effects of the fossil fuel divestment campaign on stock returns. *N. Am. J. Econ. Financ.* **2019**, *47*, 669–674. [CrossRef]

27. Sklar, M. Fonctions de repartition an dimensions et leurs marges. *Publ. Inst. Statist. Univ. Paris* **1959**, *8*, 229–231.

28. Fama, E.F.; French, K.R. A five-factor asset pricing model. *J. Financ. Econ.* **2015**, *116*, 1–22. [CrossRef]

29. Fama, E.F.; French, K.R. International tests of a five-factor asset pricing model. *J. Financ. Econ.* **2017**, *123*, 441–463. [CrossRef]

30. Hansen, B.E. Autoregressive conditional density estimation. *Int. Econ. Rev.* **1994**, *35*, 705–730. [CrossRef]

31. Patton, A.J. Modelling asymmetric exchange rate dependence. *Int. Econ. Rev.* **2006**, *47*, 527–556. [CrossRef]

32. Joe, H.; Xu, J.J. *The Estimation Method of Inference Functions for Margins For Multivariate Models*; The University of British Columbia: Vancouver, BC, Canada, 1996.

33. Breymann, W.; Dias, A.; Embrechts, P. Dependence structures for multivariate high-frequency data in finance. *Quant. Financ.* **2003**, *3*, 1–14. [CrossRef]

Article

Optimal Coordination of Wind Power and Pumped Hydro Energy Storage

Hussein M. K. Al-Masri [1,*], **Ayman Al-Quraan** [1], **Ahmad AbuElrub** [2] **and Mehrdad Ehsani** [3]

[1] Department of Electrical Power Engineering, Yarmouk University, Irbid 21163, Jordan; aymanqran@yu.edu.jo
[2] Department of Electrical Engineering, Jordan University of Science and Technology, Irbid 22110, Jordan; amabuelrub@just.edu.jo
[3] Department of Electrical and Computer Engineering, Texas A&M University, College Station, TX 77843, USA; ehsani@ece.tamu.edu
*   Correspondence: h.almasri@yu.edu.jo; Tel.: +962-721-1111-4658

Received: 29 September 2019; Accepted: 13 November 2019; Published: 19 November 2019

Abstract: A study combining wind power with pumped hydro energy storage for the Jordanian utility grid is presented. Three solvers of the Matlab optimization toolbox are used to find the optimal solution for the cost of energy in a combined on-grid system. Genetic algorithm, simulated annealing (SA), and pattern search (PS) solvers are used to find the optimal solution. The GA solution of 0.0955388 $/kWh is economically feasible. This is 28.7% lower than the electricity purchased from the conventional utility grid. The discounted payback period to recover the total cost is 10.271 years. The suggested configuration is shown to be feasible by comparing it to real measurements for this case and a previous wind-only case. It is shown that the indicators of the optimal solution are improved. For instance, carbon dioxide emissions (E_{CO2}) and conventional grid energy purchases are reduced by 24.69% and 24.68%, respectively. Moreover, it is shown that the benefits of adding hydro storage, combined with increasing the number of wind turbine units, reduces the cost of energy of renewables ($COE_{Renewables}$). Therefore, combining hydro storage with wind power is economically, environmentally, and technically a more efficient alternative to the conventional power generation.

Keywords: pumped hydro storage; wind farm; simulated annealing; genetic algorithm; pattern search; Matlab optimization toolbox; economic and environment feasibility

---

1. Introduction

Hydro and wind powers are promising renewables. However, due to the stochastic nature of the wind power, it is more efficient and reliable to combine it with another suitable energy system to provide a stable operation for large utility grid systems. Pumped hydro storage (PHS) is a suitable energy storage system that can be hybridized with wind power in order to overcome its variability and provide real-time load following. Hydro power makes up around 19% of electrical power generated worldwide [1]. It is one of the oldest methods of renewable energy generation [2]. Hydropower originates from the sun, as its water cycle is driven by solar radiation. Approximately 22% of incoming solar energy is captured to form precipitation, which is the source of hydropower [3]. Hydropower stations can be categorized based on their output power. They are classified as small, mini, or micro types when the maximum output power is 15 MW, 1 MW and 100 kW, respectively [4]. In this paper, the maximum output of the PHS exceeds the small type, therefore, a large type is added to the aforementioned category.

PHS plants are mainly used to serve demand during the peak load hours [3]. When wind generation exceeds demand, excess power can be stored by pumping water into the upper reservoir of the PHS system. Conversely, when the load exceeds the wind generation, the stored hydro energy can be used to supply the power deficit. In fact, PHS plants are considered to be one of the best utility-scale energy storage solutions due to their ability to supply power in just one to three minutes [3].

In the published literature, the operation of the grid-connected PHS, combined with wind power, has been extensively investigated. In [5], the authors suggested using pumped hydro storage as an operating reserve ancillary service in order to mitigate the problems related to wind farm integration with the grid. A probabilistic unit commitment using Lagrange relaxation was suggested to find the optimal scheduling of the thermal generators when wind power was integrated into the system while considering the uncertainty of the wind speed. It was found that pumped hydro storage could be effectively employed to reduce operating and flying reserve costs. In [6], PHS application in combination with a wind farm to increase profit in electricity markets was investigated. The results showed that the revenue was a function of the type of hydro storage used and market characteristics. The revenue increased by up to 11% by employing PHS. The authors in [7] proposed a deterministic, dynamic programming, long-term generation expansion model to find the optimal generation mix, total system cost, and total carbon dioxide emissions of a PHS system connected to a wind farm. It was found that in order to gain financial benefit from building the capital-intensive PHS, the exogenous market costs had to be very strong. In [8], a novel coordination strategy of a wind farm combined with PHS for a faster, reliable self-healing process in the grid restoration phase was proposed. The problem was formulated as a two-stage adaptive robust optimization and solved using the column-and-constraint generation (C&CG) decomposition algorithm. The results proved that the PHS could increase system reliability and reduce wind power curtailment. A combinatorial planning model in order to maximize wind power utilization and reduce wind energy curtailment was studied in [9]. A posterior multi-objective (MO) optimization approach was proposed to deal with wind energy curtailment cost and the total social cost. The obtained results introduced an optimization approach capability and efficiency regarding the planning of renewable-based power systems. In [10], a sizing method for a wind–hydro system in the Canary Islands was proposed and its economic benefits for the island's electrical system were investigated. The contribution of this wind–hydro system to satisfying electricity demand was 29% higher than wind-only, and the electrical energy generation cost was reduced by 7.68 M€/year. In [11], the authors presented an improved probabilistic production simulation method to facilitate the cost–benefit analysis of PHS. A case study on the IEEERTS79 system, which was used to demonstrate the effectiveness of the proposed simulation method, helped the industry move toward high penetration of the integrated wind energy power system.

In order for sustainable power generation to become universally adopted so that its planetary benefits are realized, the economic and technical designs of these power plants must be locally appropriate and optimal. This paper addresses this fundamental challenge for design engineers and managerial decision-makers.

The scientific/technical problem that is addressed and solved in this case study is as follows. In order to solve the global warming and cost of energy problems contributed to by electric power generation, local renewable resources must be utilized, combined, and optimized in their overall system design. This paper addresses these technical problems in a case study of combining wind and hydro power generation in Jordan as a specific location. In addition, this paper investigates the financial, environmental, and technical feasibility of wind farming and pumped hydro energy storage in an oil-importing country to reduce the energy-producing burden. Three heuristic optimization techniques are used from the Matlab optimization toolbox to verify the system design. Results show that the proposed system may be notably beneficial for Jordan. The same methodology can be applied in countries where this is relevant, such as Panama, a country which one of the authors visited for this purpose.

The fundamental problem of global sustainable energy production is the optimal use of locally specific renewable energy sources, such as wind and hydro energy resources, as laid out in this paper. In other words, this global problem must be solved locally everywhere. This is an engineering design optimization, which usually requires hybrid power plants. This paper presents a detailed case study of how this engineering problem is solved. In the process, it also sheds light on our concept of local solutions to a global problem. This important concept is often lost on countries and companies that attempt to build sustainable power generation projects.

In this paper, the Matlab optimization toolbox was used to find the optimal solution in terms of technical, environmental, and economic considerations. Moreover, genetic algorithm (GA), simulated annealing (SA), and pattern search (PS) techniques were used from the above toolbox to solve the problem described in this paper. Furthermore, it was shown that the objective function, cost of energy, of the on-grid, which was penetrated by the hydro–wind system (COE$_{PS}$) was optimally minimized. The economically feasible solution was considered to find detailed solutions. This work aims to help decision-makers find the best technical solutions before actual implementation of the proposed energy configuration.

2. Description of the Proposed System

This paper discusses the combination of a wind and hydropower system (See Figure 1), which is integrated with the distribution grid in the country of Jordan, as a case study in an oil-importing country.

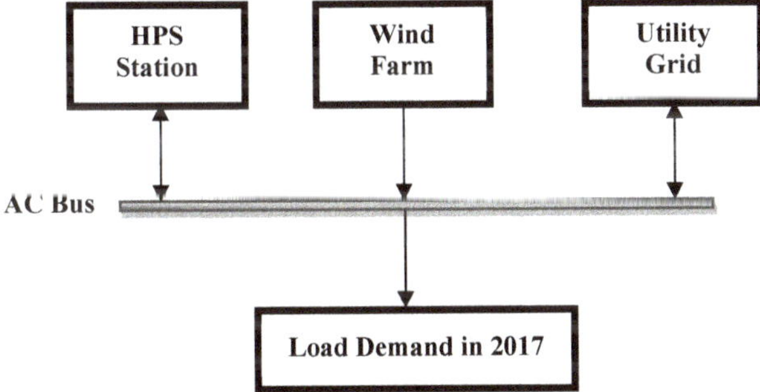

Figure 1. On-grid hydro–wind energy schematic.

The location was the same as one investigated in [12], where an on-grid wind power system was studied in Aqaba, Jordan. However, in this paper, an on-grid wind farm combined with a PHS station was investigated. Therefore, some data are the same, while others are updated for this more up-to-date study. The location was considered to be geographically suitable to construct a PHS station.

Artificial intelligence techniques (GA, SA, and PS) provided by the Matlab optimization toolbox were used to find the optimal solution of the objective function (COE$_{PS}$). Then, based on the best fitness, many indicating corresponding functions were computed, such as the wind and hydro fraction (WH$_f$), grid purchases, the footprint of the renewables, and carbon dioxide emissions (E$_{CO2}$). This procedure aimed to help design engineers replicate the same criteria to find optimal solutions for other system configurations to be adopted based on these technical studies and negotiations between electric utilities and investors. Economic, technical, and environmental feasibility impacts were also studied.

2.1. PHS Station Data

The information that was specified for the pumped hydro storage plant to be accurately modeled is shown in Table 1. First, the roundtrip efficiency referred to the ratio of the energy out to the energy in over a period of time [13]. It is difficult to separately measure the charging and discharging energies, therefore, manufacturers usually determine the round-trip efficiency and consider it to be the charging efficiency by assuming 100% discharging efficiency. Many authors have discussed this issue in the case of battery systems. Thus, the charging efficiency was set to be equal to the round-trip efficiency, and the discharging efficiency was assumed be in agreement in [14,15].

Table 1. Values of parameters used for the pumped hydro storage (PHS) station.

Parameter	Unit	Value
Lifetime	Years	50 [16]
Usable state of charge [1]	%	85
Roundtrip efficiency (ζ)	%	85 [17]
Capital cost	$/kW	1651.04
Operation and maintenance cost (OMC)	(%/Year of capital cost (CC))	1.5
Gross head	m	50
Mean water depth	m	15

Second, the usable state of charge (SOC) [1] referred to the ratio of the usable energy that was taken to the total energy of the PHS. In other words, the usable SOC was the energy left in the upper reservoir compared with the amount of energy in a full reservoir. This gave an indication into how long the PHS station could provide energy before a refill. In this study, it was assumed that a minimum stored energy should remain, and this value was the complement of the usable SOC. The usable SOC was assumed to be the same as the round-trip efficiency. Third, an initial PHS stored energy in the upper reservoir was assumed [18]. The aforementioned parameters helped to determine the PHS power generation capacity in kW, which could be used to supply the load as needed. This capacity value was sized using the GA, SA, and PS of the Matlab optimization toolbox. The capital cost of the PHS had an average value of 1651.04 $/kW [16]. The operation and maintenance costs (OMC) were taken as percentages of the capital cost (CC) [16]. Table 1 shows the values that were assumed and considered for the PHS plant. These plant data were used to compute the hourly energy generated. There was an approximate ratio of ten between the rated power (in kW) and energy (in kWh) of the PHS station, as stated in [19].

2.2. Wind Speed and Probability Distribution Function

Wind speed can change rapidly in any region. Its variation depends on several factors, such as the surface and the local weather. Appropriate predictions of wind speed in a specific area are necessary for wind power and energy estimations in that area. One of the models for characterizing the wind power is a cubic function of the wind speed. Therefore, a small error in the prediction of wind speed leads to huge variations in the wind energy estimation. Various methods are used to study the characteristics of wind speed. Weibull and Rayleigh distributions are the most preferred methods, as they are flexible and easy in terms of parameter determination.

The focus in this paper was on the Rayleigh distribution, which is a special form of the Weibull distribution with a shape factor that is always equal to two. In the Rayleigh distribution, the mean wind speed is sufficient to determine the wind characteristics. The Rayleigh distribution function ($f_R(v)$) is given by Equation (1) [20,21].

$$f_R(v) = \left(\frac{\pi}{2}\right)\left(\frac{v}{v_a^2}\right) exp - \left[\frac{\pi}{4}\left(\frac{v}{v_a}\right)^2\right]$$

(1)

where v_a is the average wind speed in a specific area in (m/s). The wind speed logarithmic law shown in Equation (2) was used to model the variation of wind speed due to the difference in height between the anemometers of the metrological station and the hub of the proposed wind turbine. In addition, it considered the terrain roughness between two altitudes [12,22].

$$\frac{v}{v_0} = \frac{ln(H/z_0)}{ln(H_0/z_0)}$$

(2)

where v_0 is the wind speed corresponding to the height (H_0) and Z_0 is the roughness coefficient. A case study was conducted in Aqaba, which is the free Trade Area in Jordan. The wind speed was measured in a specific location using anemometer installed at 45 m above ground level, in which the output data was taken on a monthly average basis. Then, Rayleigh distribution was used to obtain hourly data, as shown in Figure 2. The roughness factor of the logarithm used for this case was 0.03 to adjust for the wind speed of open terrain areas [22]. Also, the hub height of the proposed wind turbine was 80 m (Table 2) which was also considered in the logarithm. The wind speed-based Rayleigh distribution function in Aqaba for twelve months is shown in Figure 2.

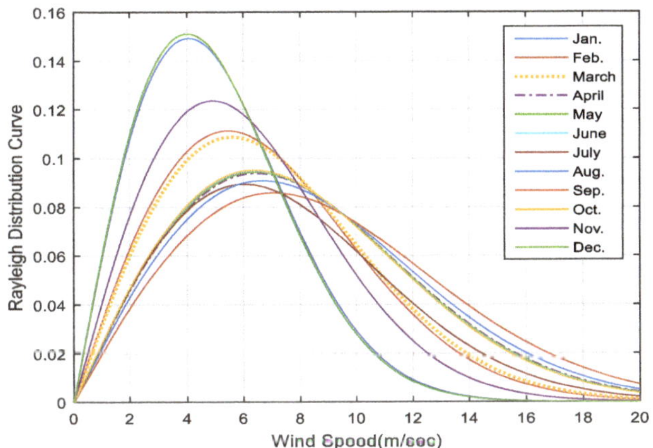

Figure 2. Curves of Rayleigh distribution for all months.

Table 2. Manufacturer information of a C96–2.5 MW wind turbine.

Turbine	
Manufacturer	USA
Power	
Rated power	2.5 MW
Cut-in wind speed	4.0 m/s
Rated wind speed	12.5 m/s
Cut-out wind speed	25.0 m/s
Survival wind speed	70.0 m/s
Rotor	
Diameter	96.0 m
Swept area	7238 m^2
Number of blades	3
Maximum rotor speed	15.5 U/min
Tip speed	78 m/s
Type	46.7
Material	Fiberglass
Power density	345.4 W/m^2
Gearbox	
Type	Spur
Stages	2.0
Tower	
Hub height	80.0 m
Type	Steel tube
Shape	conical
Corrosion protection	painted

Other information that was determined for the wind farm to be precisely sized is shown in Table 3. The financial input parameters were the same as the ones described in the wind-only investigation [12]. The project lifetime was assumed to be 50 years. Therefore, the wind turbine will be replaced twice, with a cost that was assumed the same as the capital cost.

Table 3. Lifetime and spacing parameters of the wind farm.

Parameter	Unit	Value
Life-time per unit	Years	20
Row spacing of the farm (RS)	m	384
Column spacing of the farm (CS)	m	672

The geographical area of the wind farm (A_{WF}) was computed using Equation (3). L and W are the dimensions of the wind farm, which was considered to have a rectangular shape. For the row spacing (RS) and column spacing (CS) values shown in Table 3, Equations (4) and (5) were used to calculate L and W.

$$A_{WF} = L \times W \tag{3}$$

$$L = C_S(N_{col} - 1) + D_r \tag{4}$$

$$W = R_S(N_{row} - 1) + D_r \tag{5}$$

where D_r, N_{row}, and N_{col} are the rotor diameter, number of rows, and number of columns, respectively. These helped to compute the maximum and minimum wind areas, i.e., the A_{max} and A_{min}. A footprint cap limit of 20,000 Dunam was considered for the on-grid wind hydro energy system.

2.3. Load Demand Hourly Data

The load demand hourly values of Aqaba, Jordan in 2017 were prepared after tailoring the supervisory control and data acquisition (SCADA) demand values in 2016 used in [12]. They were obtained from the National Control Center of the National Electric Power Company, Jordan. A percentage growth of 6% for a year is usually used in electric utilities in Jordan to obtain the annual load demand for the following year, therefore, in this paper, the hourly load values in 2017, as shown in Figure 3, were obtained by applying this percentage.

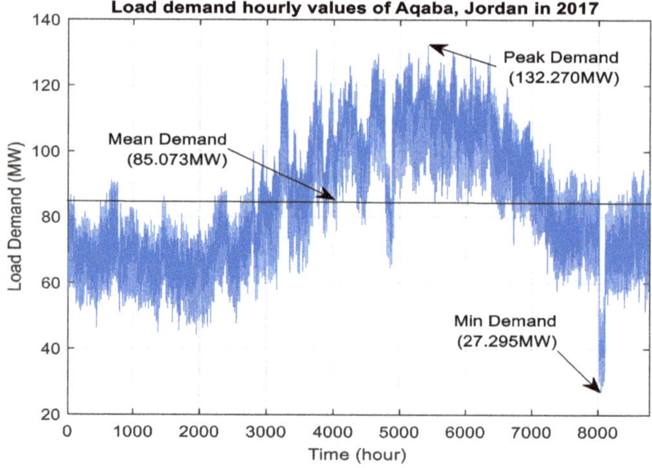

Figure 3. Load demand hourly values of the Aqaba Qasabah district in 2017.

The minimum, maximum, and mean load demand values were 27.295 MW, 132.270 MW, and 85.073 MW, respectively, as shown in Figure 3.

3. Mathematical System Formulation

3.1. Modeling of the Hydro Station

The priority was to satisfy the load from the wind farm. If the wind power was not sufficient, then energy deficit should be covered by the PHS station and, lastly, the energy purchased from the utility grid. The Matlab code had the target of satisfying the entire load. Three cases were considered. First, the load was satisfied by the wind farm, and if there was excess wind power and the reservoir was full, the generation of the hybrid renewable energy system came only from the wind power plant. Second, the load was satisfied by the wind farm, and there was excess wind power and the reservoir was not full. Thus, we computed the excess wind power that could charge the PHS plant by comparing the excess wind power value to the rated capacity of the PHS plant.

Third, when the generation of wind farm was less than what was required by the load demand, we checked the availability of the PHS plant for this power deficit. Moreover, the PHS minimum energy storage capacity was set, which was not exceeded during the discharge.

Once the rated power of the hydro station, P_{rated}, in kW was optimized, the energy in kWh, W_{rated}, was estimated based on the assumption made in Section 2.1. Then, the potential energy (in J/m^3), W_J, and (in kWh/m^3), W_{kWh}, of water in the upper reservoir were computed using Equation (6) and Equation (7), respectively.

$$W_J = \rho_{water} g H \tag{6}$$

$$W_{kWh} = 2.78 \times 10^{-7} W_J \tag{7}$$

where ρ_{water} is the density of water (1000 kg/m^3), g is the gravitational acceleration (9.81 m/s^2), and H is the actual head of the PHS station [12].

Then, the volume of the water in the upper reservoir (in m^3), V_{water}, was computed using Equation (8). At this point, the area required for the PHS station, A_{PHS}, was computed using Equation (9) for a given mean depth, D [12]. Furthermore, Equation (10) was used to compute the water flow (F_{water}) in the pipeline in (m^3/s) [23].

$$V_{water} = \frac{E_{rated}}{\eta W_{kWh}} \tag{8}$$

$$A_{PHS} = \frac{V_{water}}{D} \tag{9}$$

$$F_{water} = \frac{P_{rated}}{\eta H g} \tag{10}$$

3.2. Modeling of the Wind Turbine Power Curve

The wind turbine output power model can be typically presented in two main regions. Region 1 exists between the cut-in speed [1] and the rated wind speed (V_R), while Region 2 exists between V_R and the cut-out wind speed [6], as shown in Figure 4. This shows the ideal model representation of a wind turbine and the corresponding main regions.

To convert the hourly wind speed values, obtained before using Rayleigh distribution, into hourly output wind turbine values, the mathematical model in Equation (11) is used to model Region 1 shown in Figure 4. P_R is the rated power generated by a wind turbine. Further, the corresponding A, B and C parameters are given in Equations (12)–(14) [9,24,25]. This model is different from the ones described in [12]. The output power in Region 1 runs smoothly between V_I and V_R with no protrusions at the cut-in value, as shown in the models described in [12], see Figure 5. This will result in an accurate computation of the output power extracted from the wind farm. This leads to precise computations in the output wind power and energy and thus in the number of units sizing, geographical footprint, economic and environmental indicators.

$$P(v) = \left\{ \begin{array}{ll} P_R\big(A + Bv + Cv^2\big), & V_I \leq v \leq V_R \\ P_R, & V_R \leq v \leq V_o \end{array} \right\} \tag{11}$$

$$A = \frac{1}{(V_I - V_R)^2}\left[V_I(V_I + V_R) - 4V_I V_R \left(\frac{V_I + V_R}{2V_R}\right)^3 \right] \tag{12}$$

$$B = \frac{1}{(V_I - V_R)^2}\left[4(V_I + V_R)\left(\frac{V_I + V_R}{2V_R}\right)^3 - (3V_I + V_R) \right] \tag{13}$$

$$C = \frac{1}{(V_I - V_R)^2}\left[2 - 4\left(\frac{V_I + V_R}{2V_R}\right)^3 \right] \tag{14}$$

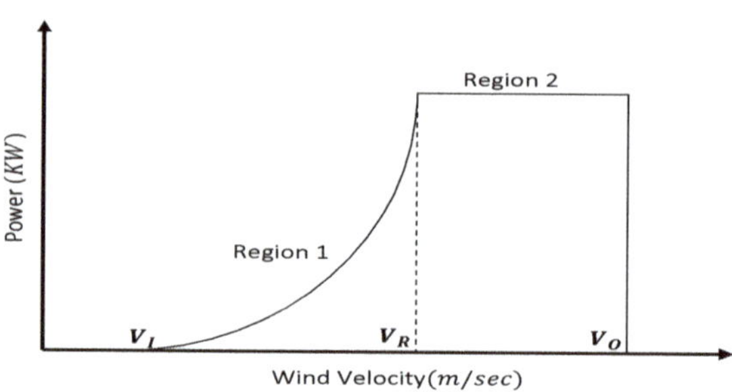

Figure 4. Schematic of wind turbine output power model.

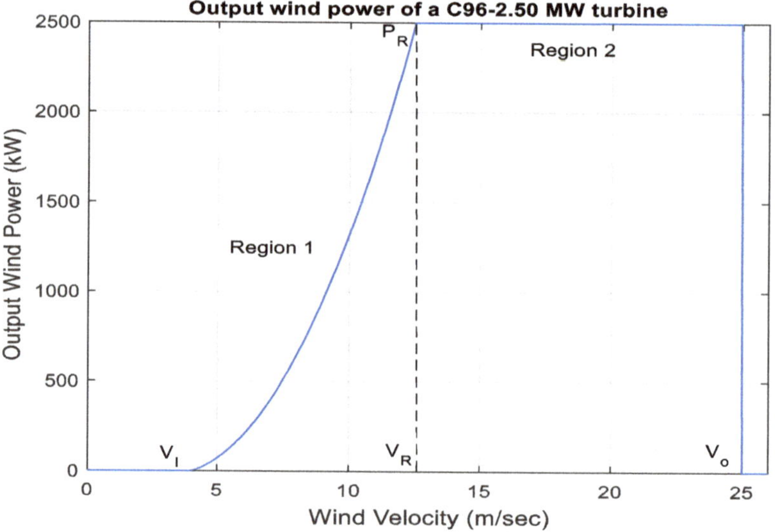

Figure 5. Output wind power of a C96–2.50 MW wind turbine.

3.3. Objective Function

The objective function in this study was selected to be the cost of energy of the penetrated system (COE$_{PS}$) to reflect the price of the energy supplied by the on-grid hybrid wind hydropower system, as designed and sized to cover the load demand shown in Figure 3. It was computed by dividing the system's cost by the system's absorbed energy, as shown in Equation (15) [26].

$$COE_{PS} = \frac{System's\ cost}{System's\ absorbed\ energy} \tag{15}$$

The system cost was computed by first constructing nominal and discounted cost cash flows for the project lifetime, i.e., 50 years. The nominal cash flow included the capital cost (CC), the replacement cost [22], the operation and maintenance cost (OMC), and the salvage cost (SC). These costs were discounted for the present in the discounted cash flow that represented the total current cost (TCC) of the renewable power system. The system cost calculation criteria are described in detail in [26]. The wind farm financial parameters are described in detail in [12]. The cost parameters of the PHS station are given in Table 1. The cost of the energy from the utility grid had the last priority to satisfy the load demand.

Thus, an economic comparison with one hour time steps was done to satisfy the load demand in Figure 3. However, there were priorities built into the design code to satisfy the load demand; the wind farm first, then the PHS plant, and, lastly, the necessity to purchase energy from the grid if the load was still not satisfied. Moreover, the number of wind turbines (WTs) was computed based on the rated power, as shown Figure 5.

3.4. Indicators of the Objective Function

There are technical, economic, and environmental indicators that were computed based on the optimal value of the COE_{PS}. These included the wind and hydro fraction (WH_f), as shown in Equation (16), and carbon dioxide emissions (E_{CO2}).

$$WH_f = \frac{Renewable\ generation}{System's\ absorbed\ energy} \tag{16}$$

The E_{CO2} was computed by summing up the hourly multiplied grid energy purchases with a grid emission factor of 583.866667 gCO_2/Wh.

4. Optimization Toolbox of Matlab

The optimization toolbox in Matlab is a collection of functions that implement Matlab's numerical capability and computing environment. This toolbox provides functions to find parameters which minimize or maximize objectives to satisfy specific constraints. Therefore, the optimal solutions of continuous and discrete problems can be obtained, tradeoff analyses can be achieved, and optimization design tasks can be performed using this toolbox. In addition, parameter estimations and tuning can be done using this toolbox. Moreover, solvers for linear programming (LP), quadratic programming, nonlinear programming (NLP), constrained linear least squares, nonlinear least squares, and nonlinear equations are included in this optimization tool box [27]. In this paper, the genetic algorithm (GA), simulated annealing (SA), and pattern search (PS) optimization methods were used from the Matlab toolbox.

GA, which is a search technique based on a principle of biological genetics and natural selection, allows a composition of many individuals to evolve under specified selection rules to a state that maximizes fitness under a specific objective function.

As a Matlab tool, GA is a powerful tool capable of providing robust approximation for systems that may be subject to uncertainties [28,29]. Its research mechanism consists of the use of candidate solutions represented in a binary form, called chromosomes. Several genetic operators, such as crossover, mutation, and inversion, are used to adapt and fit the generated population of chromosomes in each research step [29].

The flow of the genetic algorithm can be summarized by the following steps [30].

- Create initial population (usually a randomly generated string);
- Evaluate all the individuals (apply some function or formula to the individuals);
- Select a new population from the old population based on the fitness of the individuals and the required objective function;
- Apply some genetic operators (mutation, crossover, and inversion) to the population members to create new individuals;
- Evaluate the newly created individuals based on the required objective function.

Repeat the last three steps until the stopping criteria has been satisfied, where a certain fixed number of generations is obtained.

In summary, the GA toolbox has four main modules: The optimization problem definition module, the variables setting module, the generation of the initial population module, and the evolution module. These modules interact with each other by exchanging information that enables the operation of the algorithm. Before running the optimization algorithm, it is necessary to characterize the optimization problem. Then, the type and the representation of the variables used by the algorithm must be defined. GA works directly with real variables or with codified variables. Thus, depending on the type of variable defined by the problem and the type of representation used by the GA, there is a necessity for coding/decoding to pass from the actual workspace to the GA workspace.

Moreover, pattern search (PS), i.e., direct search or derivative-free search, is one of the Matlab optimization methods used to optimize functions that are not continuous or differentiable. Optimization attempts to find the best-match solution with the lowest error value in a multidimensional analysis space of possibilities [27]. Furthermore, simulated annealing (SA) is a Matlab toolbox method used to solve unconstrained and constrained optimization problems [31,32]. The models of this method simulate the heating process of the materials. At each iteration step of the simulated annealing algorithm, a new point is randomly generated. The distance of the new point from the current point is based on a probability distribution with a scale proportional to the temperature. An annealing schedule is selected to systematically decrease the temperature as the algorithm proceeds. As the temperature decreases, the algorithm extends its search to finally reach an optimal solution. The SA algorithm consists of two main options, namely, "AcceptanceFcn" and "TemperatureFcn". The first option accepts the worst case in order to achieve a global solution for the desired problem. The second option selects the suitable algorithm uses to update the temperature. Two stopping criteria are used for the SA algorithm, which are function tolerance and maximum iterations. In the first criterion, the algorithm runs until the average change in value of the objective function is less than the value of tolerance. In the second criterion, the maximum number of iterations can be determined [27].

5. Results and Discussion

Every component shown in Figure 1 was modeled and coded in Matlab along with the objective function of the cost of energy of penetrated system (COE$_{PS}$) and the rest of the corresponding indicators. Table 4 shows the results obtained using the GA, SA, and PS solvers. Also, many data corresponding to the optimal value of the COE$_{PS}$ are included in Table 4. The three aforementioned solvers of the Matlab optimization toolbox were selected to solve the problem described in this paper. The SA and PS solvers provided solutions that were 1.27634% and 1.98903% higher than the GA solution, respectively. Therefore, the GA solution was found to be feasible compared with the other solutions.

Table 4. Detailed results of the optimized system using genetic algorithm (GA), simulated annealing (SA), and pattern search (PS) algorithms.

Parameter	Value (GA)	Value (SA)	Value (PS)
COE_{PS} ($/kWh)	0.0955388	0.0967582	0.0974391
E_{CO2} (kt/year)	198.9044	204.5988	207.5809
$COE_{Renewables}$ ($/kWh)	0.0631	0.0637	0.0641
Number of WTs	47	45	44
A_{max} (m^2)	9,676,800	7,142,400	8,506,368
A_{max} (Dunam)	9676.800	7142.400	8506.368
A_{min} (m^2)	6,856,704	6,561,792	6,266,880
A_{min} (Dunam)	6856.704	6561.762	6266.880
Total cost (M$)	441.3	426.37	419.44
WH_f (%)	56.1427	54.3791	53.4641
Grid purchases cost (M$)	45.649	46.956	47.641
Grid energy purchases (GWh)	340.67	350.42	355.53
P_{rated} (PHS) (kW)	18,118.5	19,226.37	20,039.45
E_{rated} (PHS) (kWh)	181,185	192,263.7	200,394.5
W_J (J/m^3)		4.905×10^5	
W_{kWh} (kWh/m^3)		0.136359	
V_{water} (m^3)	1.56322×10^6	1.65880×10^6	1.72895×10^6
A_{PHS} (in m^2)	104,214.902	110,586.803	115,263.501
A_{PHS} (in Dunam)	104.2149	110.5868	115.2635
F_{water} (m^3/s)	43.457	46.115	48.065

The GA solution of 0.0955388 $/kWh was economically feasible compared with the SA and PS COE_{PS} values. The optimal value of the COE_{PS}, which was found using the GA, is shown in Figure 6. This value was 28.7% less than the energy bought from the conventional electric network, which is an excellent indication for the economic feasibility of this suggested configuration.

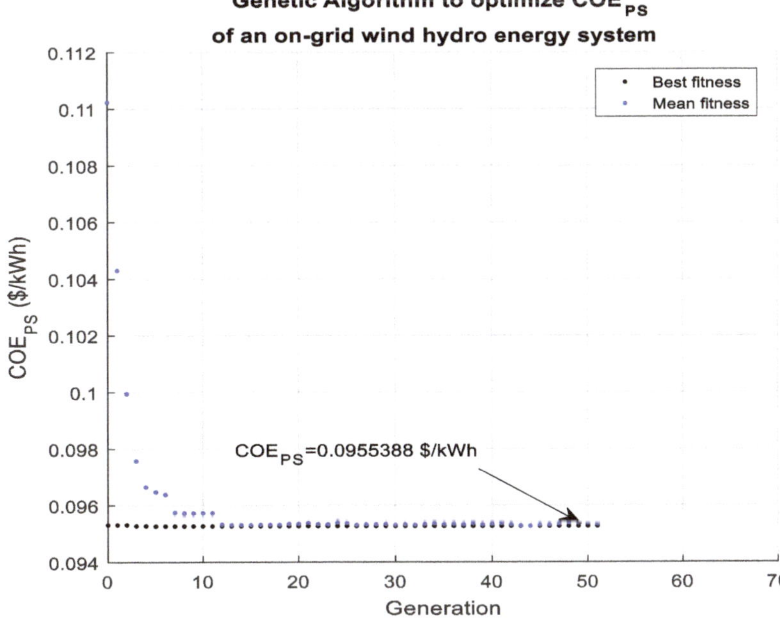

Figure 6. GA to optimize COE_{PS}.

Further, Figure 7 shows the current best point for the two decision variables found at the optimal value of the objective function.

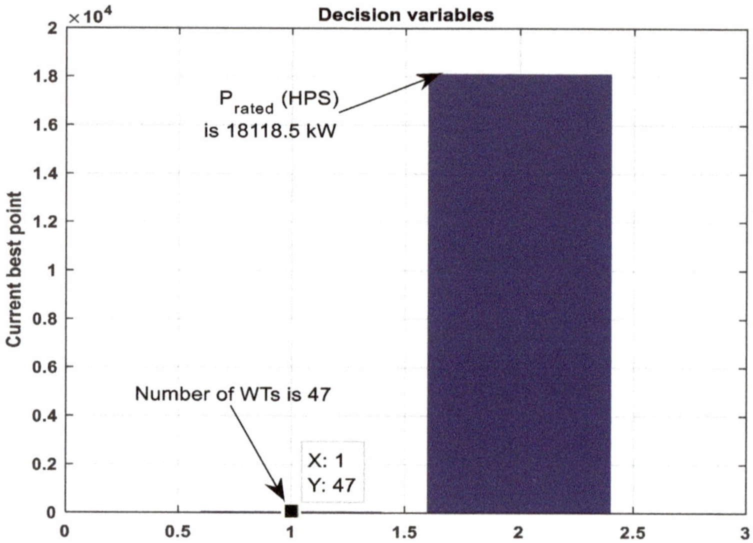

Figure 7. Corresponding decision variables of the optimal configuration.

Note that the area of the wind farm was assumed to be rectangular and was therefore computed by incrementing the odd optimal number by one. Furthermore, the E_{CO2} in the suggested location was 634.645 kt/year [12], therefore, the emissions were mitigated by 68.66%, assuming that the renewable configuration in Figure 1 was adopted. Also, the geographical area of renewable plants ($A_{Renewables}$) was increased. However, only 48.91% of the geographical area limit was used to install the designed hydro-wind energy system. Thus, the rest of the area (51.09%) could be used in the future as load demand and the system size grow.

The discounted payback period (DPP) is frequently used in renewable energy studies to find the length of time needed to retrieve the initial investment [33–35]. This was done in this paper by building the cumulative cash flow (CCF), as shown in Table 5. Note that the present value factor (PVF), cash flow (CF), and the corresponding discounted cost values ($CF_{discounted}$) were calculated. The CF in Table 5 included the total cost found before in Table 4 using GA, and the energy savings of the renewable energy system. These energy savings were computed by multiplying the yearly renewable generation (436.438 GWh) by the energy purchased price of electric utilities in Jordan. Afterward, the CCF values were computed by cumulatively adding the discounted cost values. Then, the time to get back the total cost value was calculated using Equation (17). Note that Table 5 shows only 15 years of the 50-year project life-time, because the aim was to obtain a positive cost value from the CCF, which was held at the 11th year. This was just before the time when the total cost was retrieved. Table 5 shows that the DPP was computed to be around 10.271 years (10 years, 3 months, and 7 days).

$$DPP = n_l + \frac{|CCF\,(n_l)|}{CF_{discounted}\,(n_l + 1)} \tag{17}$$

where n_l is the year number at the last negative cost value of the CCF.

Table 5. Calculation of the discounted payback period.

Year (N)	PVF	CF (M\$)	CF$_{discounted}$ (M\$)	CCF (M\$)
0	1.000	−441.300	−441.300	−441.300
1	0.944	58.483	55.234	−386.066
2	0.892	58.483	52.165	−333.901
3	0.842	58.483	49.267	−284.633
4	0.796	58.483	46.530	−238.103
5	0.751	58.483	43.945	−194.158
6	0.710	58.483	41.504	−152.654
7	0.670	58.483	39.198	−113.456
8	0.633	58.483	37.020	−76.436
9	0.598	58.483	34.964	−41.472
10	0.565	58.483	33.021	−8.451
11	0.533	58.483	31.187	22.736
12	0.504	58.483	29.454	52.190
13	0.476	58.483	27.818	80.008
14	0.449	58.483	26.272	106.280
15	0.424	58.483	24.813	131.093
DPP (Years)			10.27096862	

The study performed in this paper, after adding the hydro storage system, was compared to a previously studied scenario in [12] for a wind-only system connected to the utility grid at the same location. Table 6 shows the percentage increase/decrease for the parameters computed in Table 4. For the wind–hydro on-grid system, the COE$_{PS}$ and the grid purchases were reduced by 16.93% and 24.68%, respectively, showing the importance of the storage system for wind power that fluctuates naturally. These cost and emissions reductions are significant, especially for non-oil producing countries, such as Jordan, which imports around 96% of its energy needs as oil and natural gas. The carbon emissions reduction was improved compared with the wind-only system. Furthermore, renewable penetration increased by 56.64% as a result of adding the PHS system, resulting in a more environmentally friendly power system.

Table 6. On-grid wind farm with/without PHS comparison.

Parameter	Percentage Increase (+) or Decrease (−) in %
COE$_{PS}$	−12.26
E$_{CO2}$	−24.69
COE$_{Renewables}$	−1.52
Number of WTs	+104.35
A$_{Renewables}$ (max)	+71.14
A$_{Renewables}$ (min)	+21.79
Total cost	+110.83
WH$_f$	+56.64
Grid purchases cost	−24.69
Grid energy purchases	−24.68

6. Conclusions

In this paper, every component shown in Figure 1 was modeled and coded in Matlab along with the objective function (COE$_{PS}$). A wind–hydro grid connected power system was proposed as an adjunct to an existing power grid. This was mathematically modeled and then coded in Matlab. The GA of the Matlab optimization toolbox was used to find the optimal feasible value of the COE$_{PS}$, which was 0.0955388 \$/kWh. This was 28.7% less than the conventional energy from the power grid. The discounted payback period was 10 years, 3 months, and 7 days. Furthermore, carbon emissions were reduced by 68.66% compared with experimentally estimated data. As a result, the grid energy purchases were also reduced. Specifically, comparing the system described in this study with the formerly studied on-grid wind-only system showed that the COE_{PS}, E_{CO2}, $COE_{Renewables}$, and grid energy purchases were reduced by 12.26%, 24.69%, 1.52%, and 24.68%, respectively. These are very promising results, especially for oil-importing countries, such as Jordan, where imported energy is a significant financial burden to the economy. The proposed wind power system with hydro storage is recommendable for its clean and economical features, compared with the conventional fossil-fueled grid or wind-only on-grid renewable configurations.

Finally, this paper is a case study to demonstrate the important point of local solutions to the global problem of global warming. The paper is necessarily limited to the specific data and assumptions of the local case study. Future work will include applying the above principle and the methodology of this paper to many other local engineering boundary conditions.

Author Contributions: H.M.K.A.-M. wrote the paper, brought the measured data from formal institutions in Jordan, performed the simulation and optimization using Matlab optimization toolbox, mathematically modeled the system and coded these models in Matlab; M.E. helped obtain more meaningful results and reviewed the article before the final submission; A.A.-Q. reviewed the paper and performed the pdf distribution to convert the monthly average wind speed measured data into hourly data; A.A. reviewed the paper and wrote the literature review part.

Funding: This research received no external funding.

Acknowledgments: The authors would like to acknowledge Yarmouk University, Irbid, Jordan, for their support in this research.

Conflicts of Interest: The authors declare no conflict of interest.

References

1. Bostan, I.; Gheorghe, A.V.; Dulgheru, V.; Sobor, I.; Bostan, V.; Sochirean, A. *Resilient Energy Systems: Renewables: Wind, Solar, Hydro*; Springer Science & Business Media: Berlin/Heidelberg, Germany, 2013; Volume 19.
2. Hammad, M.; Aburas, R.; Abuzahra, B. The potential of hydropower generation in Jordan. *Energy Policy* **1994**, *22*, 523–530. [CrossRef]
3. Wagner, H.-J.; Mathur, J. *Introduction to Hydro Energy Systems: Basics, Technology and Operation*; Springer Science & Business Media: Berlin/Heidelberg, Germany, 2011.
4. Fritz, J.J. *Small and Mini Hydropower Systems: Resource Assessment and Project Feasibility*; McGraw-Hill: New York, NY, USA, 1984.
5. Kiran, B.D.H.; Kumari, M.S. Demand response and pumped hydro storage scheduling for balancing wind power uncertainties: A probabilistic unit commitment approach. *Int. J. Electr. Power Energy Syst.* **2016**, *81*, 114–122. [CrossRef]
6. Castronuovo, E.D.; Usaola, J.; Bessa, R.; Matos, M.; Costa, I.; Bremermann, L.; Lugaro, J.; Kariniotakis, G. An integrated approach for optimal coordination of wind power and hydro pumping storage. *Wind Energy* **2014**, *17*, 829–852. [CrossRef]
7. Foley, A.; Leahy, P.; Li, K.; McKeogh, E.; Morrison, A. A long-term analysis of pumped hydro storage to firm wind power. *Appl. Energy* **2015**, *137*, 638–648. [CrossRef]
8. Golshani, A.; Sun, W.; Zhou, Q.; Zheng, Q.P.; Wang, J.; Qiu, F. Coordination of Wind Farm and Pumped-Storage Hydro for a Self-healing Power Grid. *IEEE Trans. Sustain. Energy* **2018**, *9*, 1910–1920. [CrossRef]
9. Hozouri, M.A.; Abbaspour, A.; Fotuhi-Firuzabad, M.; Moeini-Aghtaie, M. On the use of pumped storage for wind energy maximization in transmission-constrained power systems. *IEEE Trans. Power Syst.* **2015**, *30*, 1017–1025. [CrossRef]
10. Portero, U.; Velázquez, S.; Carta, J.A. Sizing of a wind-hydro system using a reversible hydraulic facility with seawater. A case study in the Canary Islands. *Energy Convers. Manag.* **2015**, *106*, 1251–1263. [CrossRef]
11. Zhou, B.; Liu, S.; Lu, S.; Cao, X.; Zhao, W. Cost–benefit analysis of pumped hydro storage using improved probabilistic production simulation method. *J. Eng.* **2017**, *2017*, 2146–2151. [CrossRef]
12. Al-Masri, H.M.; AbuElrub, A.; Ehsani, M. Optimization and Layout of a Wind Farm Connected to a Power Distribution System. In Proceedings of the 2018 IEEE International Conference on Industrial Technology (ICIT 2018), Lyon, France, 20–22 February 2018; pp. 1049–1054.
13. Ackermann, T. *Wind Power in Power Systems*, 2nd ed.; John Wiley: Chichester, UK, 2012.

14. Wei, Z.; Hongxing, Y.; Lin, L.; Zhaohong, F. Optimum Design of Hybrid Solar-wind-diesel Power Generation System Using Genetic Algorithm. *Hkie Trans.* **2007**, *14*, 82–89. [CrossRef]
15. Gupta, R.A.; Kumar, R.; Bansal, A.K. Economic analysis and design of stand-alone wind/photovoltaic hybrid energy system using Genetic algorithm. In Proceedings of the 2012 International Conference on Computing, Communication and Applications, Dindigul, India, 22–24 February 2012; pp. 1–6.
16. Zach, K.; Auer, H.; Lettner, G. *Facilitating Energy Storage to Allow High Penetration of Intermittent Renewable Energy*; Intelligent Energy Europe: Brussels, Belgium, 2012.
17. Schoenung, S. *Energy Storage Systems Cost Update*; SAND2011-2730; Sandia National Laboratory: Albuquerque, NM, USA, 2011.
18. Makarov, Y.V.; Du, P.; Kintner-Meyer, M.C.W.; Jin, C.; Illian, H.F. Sizing Energy Storage to Accommodate High Penetration of Variable Energy Resources. *IEEE Trans. Sustain. Energy* **2012**, *3*, 34–40. [CrossRef]
19. Rastler, D. *Electricity Energy Storage Technology Options: A White Paper Primer on Applications, Costs and Benefits*; Electric Power Research Institute: Palo Alto, CA, USA, 2010.
20. Al-Quraan, A.; Alrawashdeh, H. Correlated capacity factor strategy for yield maximization of wind turbine energy. In Proceedings of the 2018 5th International Conference on Renewable Energy: Generation and Applications (ICREGA), Al Ain, UAE, 25–28 February 2018; pp. 264–267.
21. Al-Quraan, A.A.; Pillay, P.; Stathopoulos, T. Use of a wind tunnel for urban wind power estimation. In Proceedings of the PES General Meeting Conference & Exposition, National Harbor, MD, USA, 27–31 July 2014; pp. 1–5.
22. Bañuelos-Ruedas, F.; Angeles-Camacho, C.; Rios-Marcuello, S. Methodologies used in the extrapolation of wind speed data at different heights and its impact in the wind energy resource assessment in a region. In *Wind Farm-Technical Regulations, Potential Estimation and Siting Assessment*; InTech: Vienna, Austria, 2011.
23. Gatte, M.T.; Kadhim, R.A. Hydro power. In *Energy Conservation*; InTech: Vienna, Austria, 2012.
24. Park, G.L. *Planning Manual for Utility Application of Wecs*; Michigan State Univ., Div. of Engineering Research: East Lansing, MI, USA, 1979.
25. Billinton, R.; Chen, H.; Ghajar, R. A sequential simulation technique for adequacy evaluation of generating systems including wind energy. *IEEE Trans. Energy Convers.* **1996**, *11*, 728–734. [CrossRef]
26. Singh, S.; Kaushik, S.C. Optimal sizing of grid integrated hybrid PV-biomass energy system using artificial bee colony algorithm. *IET Renew. Power Gener.* **2016**, *10*, 642–650. [CrossRef]
27. Coleman, T.; Zhang, Y. *Optimization Toolbox™ User's Guide*; Matlab Mathworks Inc.: Natick, MA, USA, 2018.
28. Haupt, R.L.; Haupt, S.E. *Practical Genetic Algorithms*, 2nd ed.; Wiley: Hoboken, NJ, USA, 2004.
29. Chaoui, H.; Khayamy, M.; Okoye, O.; Gualous, H. Simplified Speed Control of Permanent Magnet Synchronous Motors using Genetic Algorithms. *IEEE Trans. Power Electron.* **2018**, *34*, 3563–3574. [CrossRef]
30. Mohammadi, A.; Asadi, H.; Mohamed, S.; Nelson, K.; Nahavandi, S. Multiobjective and Interactive Genetic Algorithms for Weight Tuning of a Model Predictive Control-Based Motion Cueing Algorithm. *IEEE Trans. Cybern.* **2018**, *49*, 3471–3481. [CrossRef] [PubMed]
31. Ferreira, F.; Lemos, F. Unbalanced electrical distribution network reconfiguration using simulated anneling. In Proceedings of the 2010 IEEE/PES Transmission and Distribution Conference and Exposition: Latin America (T&D-LA), Sao Paulo, Brazil, 8–10 November 2010; pp. 732–737.
32. Anantathanavit, M.; Munlin, M.-A. Fusing Binary Particle Swarm Optimzation with Simulated Annealing for Knapsack Problems. In Proceedings of the 2014 9th IEEE Conference on Industrial Electronics and Applications, Hangzhou, China, 9–11 June 2014; pp. 1995–2000.
33. Jan, I. *Discounted Payback Period*. Available online: https://bit.ly/2MuZPMP (accessed on 19 October 2019).
34. Gorshkov, A.; Vatin, N.; Rymkevich, P.; Kydrevich, O. Payback period of investments in energy saving. *Mag. Civ. Eng.* **2018**, *78*, 65–75.
35. Abuelrub, A.; Saadeh, O.; Al-Masri, H. Scenario aggregation-based grid-connected photovoltaic plant design. *Sustainability* **2018**, *10*, 1275. [CrossRef]

Article

Assessing Renewable Energy Sources for Electricity (RES-E) Potential Using a CAPM-Analogous Multi-Stage Model

Paulino Martinez-Fernandez [1], Fernando deLlano-Paz [1,*], Anxo Calvo-Silvosa [1] and Isabel Soares [2,*]

1   Department of Business, Faculty of Economics and Business, University of A Coruna, Campus de Elvina, 15071 A Coruna, Spain; paulino.martinez@udc.es (P.M.-F.); calvo@udc.es (A.C.-S.)
2   Department of Economy, Faculty of Economics and CEF.UP, University of Porto, Rua Dr. Roberto Frias, 4200-464 Porto, Portugal
*   Correspondence: fdellano@udc.es (F.d.-P.); isoares@fep.up.pt (I.S.); Tel.: +34-881012599 (F.d.-P.)

Received: 31 July 2019; Accepted: 19 September 2019; Published: 20 September 2019

Abstract: Carbon mitigation is a major aim of the power-generation regulation. Renewable energy sources for electricity are essential to design a future low-carbon mix. In this work, financial Modern Portfolio Theory (MPT) is implemented to optimize the power-generation technologies portfolio. We include technological and environmental restrictions in the model. The optimization is carried out in two stages. Firstly, we minimize the cost and risk of the generation portfolio, and afterwards, we minimize its emission factor and risk. By combining these two results, we are able to draw an area which can be considered analogous to the Capital Market Line (CML) used by the Capital Asset Pricing model (CAPM). This area delimits the set of long-term power-generation portfolios that can be selected to achieve a progressive decarbonisation of the mix. This work confirms the relevant role of small hydro, offshore wind, and large hydro as preferential technologies in efficient portfolios. It is necessary to include all available renewable technologies in order to reduce the cost and the risk of the portfolio, benefiting from the diversification effect. Additionally, carbon capture and storage technologies must be available and deployed if fossil fuel technologies remain in the portfolio in a low-carbon approach.

Keywords: energy planning; Modern Portfolio Theory (MPT); Capital Asset Pricing Model (CAPM); low-carbon economy; renewable energy deployment; environmental efficiency

---

1. Introduction

Energy planning fosters decision-making from the political, social, and environmental dimensions [1]. In addition to these dimensions, physical and technical variables complete a picture characterized by the uncertainties, risks, and complexities around them [2–4]. In the core of this problem, we find some questions such as how to satisfy the demand, how to minimize the generation cost, and how to meet the emission objectives [5]. Decision-making inside energy planning is, thus, a strategic process as it conditions the economic, social, and climate future through the selected energy transition approach [6,7], so much so that the errors in energy planning can result in the overcapacity either of the total power-generation system or of some technology; in higher tariffs for the households; in higher costs for the support schemes; or in energy policy actions driving to situations of legal uncertainty for technological investors [8].

Energy planning uses several techniques to solve a territory or country energy problem. The model that we put forward in this work employs quadratic programming as it is based on the Financial Portfolio Theory to design efficient power-generation real asset portfolios [9–22]. This proposal is in the line with other well-known optimisation models such as multi-period linear programming [1,5], interval linear programming [3,23,24], or stochastic programming [2,25–27].

Energy planning can be seen as a long-term investment selection problem. The aim is to design the composition of the generation mix by complying with the economic, social, and environmental criteria set for a territory or region [10,14–17]. This approach is similar to that of the Energy Trilemma from Stempien and Chan (2017) [28], which encompasses energy security, energy sustainability, and energy equity. The proposal includes both the traditional energy planning approach about the minimisation of cost of the future power-generation portfolio that satisfies the power demand [2,4,11,29] and the current trends observed in the literature about the need of considering the risk and the uncertainty of the energy and climate context [2–4].

We employ the Modern Portfolio Theory (MPT) methodology. It allows us to include some fundamental matters in the analysis, like energy security [30–32]—through the study of the portfolio diversification—and its positive effects on the power supply disruption [12–14,17,33]. Energy security is a key element in the agenda of the energy-resource-importing countries [34]. Setting and managing their energy policy objectives are essential for these countries to reduce the energy risk arisen from importing resources [7,35]. Increasing the renewable energy sources (RES), improving the energy efficiency, and reducing CO_2 emissions [34] are objectives that empower the environmental aspect as well as improve the level of energy security. The aim is to adjust the available resource consumption and to give priority to indigenous technologies (renewable technologies) in power generation. This leads to reducing the dependence on imported fossil combustibles, to decreasing the pollutant gas emission—as these technologies are non-pollutant—and to using natural resources that are not constrained by a future depletion of reserves [5,36,37]. Thus, we can conclude that energy planning and environmental protection are two sides of the same coin [38,39].

Our approach falls within the research line aiming at including sustainable development principles inside long-term environmental and energy planning [40–42]. Examples of this environmental commitment are establishing objectives for the reduction of pollutant gases emission [14,17,31–33], for including the externalities derived from power generation [10,41–45], or for taking into account the CO_2 emission costs [10,18,29,46–48] by establishing markets like the EU Emissions Trading System (EU-ETS) [49–52]. Such a set of measures helps policy makers fix the "market failure" [44] and assign resources in a way that is optimal and efficient [41,45]. In this context, renewable technologies appear as a part of the solution due to their multiple positive features: they do not emit pollutant gases, they do not depend on fossil fuels, and thus they are not subject to geopolitical risks; as they have an autonomous character, they reduce the energy dependency [16,33,53–55].

Our proposal may improve the previous approach of Martinez et al. (2018) [56], employing and developing the initial financial concept of the Capital Market Line to a realistic power portfolio case. The process starts by characterizing the power-generation-emitting technologies according to their emission and risk. The objective is to minimize the power-generation emission level. In other words, we opted to minimize the emission amount (kg-CO_2/MWh) instead of minimizing its monetary value. This allows us to achieve a result that equalises the imbalance caused by the minimization of emission costs [57]. The model solution results in the power-generation mix that minimizes CO_2 emissions. The model consists of a multi-stage MPT model of quadratic optimization.

Among the criticism of employing the MPT to implement power-generation optimization, we find the original MPT being applied to fungible and almost infinitely divisible financial assets. As a matter of fact, power-generation plants are neither fungible nor divisible assets. We commented that the MPT, when applied to power generation, should be understood as a tool for decision-making in the medium- and long-term and for a relatively vast territory (a country or region). Under these assumptions, power-generation plants can be considered fungible and divisible assets. This study could be presented as a case of the European Union, due to the definition of technological limitations in the proposed MPT model. However, data on technological costs and emission factors belong to international institutions (IEA), which could allow us to talk about a general case and not only a European one.

It is important to highlight that the application of the MPT to power generation entails a supply approach. An optimal energy mix is clearly dependent on the demand level, and that level is the one to which the energy supply is matched. Nevertheless, the MPT approach does not deal with demand-side issues and focuses uniquely on the optimization of the power-generation mix.

Two fundamental elements are in the basis of the contribution of our work. First, this approach focuses on the environmental dimension of the power-generation portfolio. We modify the objective function, which becomes an emission-minimization function instead of a cost—or cost risk—minimization function. Hence, the model solution shows the minimum possible emission level instead of the accomplishment of the emission reduction objectives [14,17,20,32]. The second element is that the Capital Asset Pricing Model (CAPM) is employed coming from the financial field of yield optimization. The following section is dedicated to briefly explaining the MPT and the CAPM.

2. Materials and Methods

2.1. The Modern Portfolio Theory and the Capital Asset Pricing Model

According to finance theory, given a portfolio of financial assets, it is assumed that their expected yield and risk can be calculated. In the original model [57], the average of historical yields of every asset is used as an estimator of the expected yield of that asset, while the variance of those historical yields is used to assess the risk of every asset. The overall yield of a portfolio is a weighted average of the yields of its components. Likewise, the risk of the portfolio can be calculated as a quadratic weighted function of the risk of its components. MPT aims to calculate the asset participation shares that minimize the risk of the portfolio subject to certain constraints: the participation shares must sum up to one and, eventually, they must be positive, indicating that short positions are not allowed.

By solving the previous problem, we obtain the so-called efficient frontier. That is the line, in a risk-return coordinate axis, on which every efficient portfolio lies. An efficient portfolio is a portfolio that shows the minimum risk for a given return or the maximum return for a given level of risk. The efficient frontier is concave, and it represents the upper-left limit of the feasible set: the part of the plane on which every combination of risk and return lies.

MPT has not given the "best" portfolio among the efficient portfolios yet. If now we assume that there is a riskless asset, an investor can spend part of their budget in this riskless asset and the other part in an efficient portfolio [58]. The set of possible linear combinations define a line that connects the point corresponding to the riskless asset yield (in the ordinate axis as its risk is zero) and the efficient frontier. As the efficient frontier is concave, this line will be tangent to it. The tangency point is known as the tangent portfolio or the market portfolio and represents the efficient portfolio that best summarizes the market behaviour.

This tangent line is called the "Capital Market Line" or CML, and every investor should choose a point on that line as it shows higher yields for any level of risk. In fact, this line can be drawn further from the tangency point, indicating that the investor is borrowing money. The CML is a relevant concept of the Capital Asset Pricing Model or CAPM [59]. Another important issue of the CAPM focuses on a single asset in order to calculate its "Security Market Line" or SML. The slope of this SML, on a market risk–return coordinate axis, defines how the asset behaves in relation to the market. The slope of the SML is known as the beta of an asset. If the beta is less than one, the asset is less volatile (less risky) than the market. If it is higher than one, the asset is more volatile (riskier) than the market.

2.2. Applying MPT and CAPM Beyond Finance

The portfolio optimization approach can be classified inside risk-control management [60] and focuses on minimizing the portfolio risk by its diversification [16,19,60,61]. A significant number of works in the literature support the conclusion that it is a proven methodology for its application to energy planning [16,17] and for optimizing the operation of demand response resources [62]. The MPT approach allows the joint inclusion of both cost and cost-risk of the different available technologies. This duality allows that the objective function can be expressed both as a power-generation portfolio cost minimization function and as a power-generation portfolio risk minimization function. By introducing the binomial cost-risk in the energy planning optimization model, the approach (traditionally based on the cost minimization [11,13,29]) is improved.

MPT applications are extensive in the literature. Particularly, it has been helpful to analyse different environmental elements. Among the last works published about MPT applied to biodiversity, the one from Yemshanov et al. (2014) [63] stands out. Their work deals with the pest surveillance problem from a diversification point of view, and they study the optimal allocation of surveillance resources by employing MPT. Later, Akter et al. (2015) [40] applies MPT analysis to asset-based biosecurity decision support. These works underline the value of MPT as a valid and relevant tool against the uncertainty derived from the lack of knowledge about species invasion dynamics. These authors' approach is the opposite of the one maintained by others like those of References [31,64–66], who point out that, in contexts characterized by ignorance and uncertainty about the analysed assets, using historical data as the only source to develop the MPT model can drive to non-robust results. To solve this pitfall, Hickey et al. [31] employed, complementarily to MPT, other tools like diversification indexes or an approach based in real options.

Regarding CAPM, recent studies applied it to the energy field. It is worth highlighting some of these studies. Inside the power retailer portfolios management, Charwand et al. (2017) [67] studied the maximization under uncertainty of the total expected rate of return of an electricity retailer. They broadened the work of Karandikar et al. (2007) [68] who used CAPM to determine the retail electricity price for end users. In this line, Rohlfs and Madlener (2013) [69] applied CAPM to calculate every technology rate of return inside a stochastic NPV approach—they proposed a cost effectiveness model to analyse different clean-coal technology pathways from the value of capture-readiness. Other authors [70–72] confirmed the suitability of using discount models as assessment tools when valuing investment projects under conditions of risk and uncertainty. In these models, one of the key variables is the discount rate and the CAPM arises as an optimal tool for its estimation. Recently, Zhang and Du (2017) [73] referred to the work of Broadstock et al. (2014) [74] as an example of CAPM application—in this case, to investigate the possible relationship between the international oil prices and the energy stock quotes in China. In a similar line, Schaeffer et al. (2012) [75] used CAPM to study the evolution of different oil companies' stock prices and to estimate their beta, which allows to foresee the behaviour of every company in the face of changes in the market portfolio. Additionally, Mo et al. (2012) [76] put forward a multifactor market model based on the CAPM theory to study the impact of the EU-ETS on the corporate value of EU electricity firms.

2.3. Developing the Multi-Stage Model

Throughout this subsection, we are going to explain how to develop our model. This part is divided in three main steps. In the first one, we will explain how to devise the instrumental model, which considers all the power-generation technologies in order to draw a reference efficient frontier. In the second step, we deal with non-pollutant power-generation technologies—nuclear, onshore wind, offshore wind, hydro, small hydro and photovoltaic (PV)— to build a non-pollutant efficient frontier. These two first steps constitute the first stage of our model.

Finally, in the third step, we will put forward the emission-risk model based on the CO_2 emissions of the pollutant technologies and in the risk or variability of the CO_2 emission cost. This will give us a pollutant-technology-efficient frontier which corresponds to the efficient-pollutant-generation portfolios—those that offer the lowest emission level to a given level of risk or vice versa—that are to be combined with the efficient non-pollutant generation portfolios obtained in the first stage of the model.

2.3.1. The Cost-Risk Instrumental Model

Based on MPT, we devise a model to find the efficient frontier or the set of portfolio cost-risk pairs that offers the lowest cost for a given level of risk or the lowest risk for a given level of cost. We will work with twelve technologies: Six of them are non-pollutant—nuclear, onshore and offshore wind, large and small hydro, and large photovoltaic (PV)—and the other six are pollutant—coal, coal with carbon capture and storage (CCS), natural gas, natural gas with CCS, oil, and biomass.

In MPT, the cost risk of a specific technology is measured by its cost standard deviation. Table 1 shows the expected costs, the cost standard deviation, the expected emission factor, and the emission cost standard deviation for every technology in our model. We used some information available in the literature about the different categories of costs in a power-generation plant (capital expenditures and operational expenditures, such as operation and maintenance costs, fuel costs, emission costs, and dismantling costs) to calculate the average generation costs by technology, their standard deviation, and the correlation among them. We also use the emission cost standard deviation as a proxy of the real emission standard deviation, as we have no real emission data. It is important to underline that we consider the nuclear generation technology as non-pollutant, even though it involves other important environmental risks not related with carbon emission. Moreover, we decided to include biomass generation in the pollutant set of technologies as it has carbon emissions, although it could be considered testimonial. In fact, this decision (considering the biomass as a pollutant technology) will have some effects on our results that will be conveniently explained. Another point to take into account is the fact that the current emission factor can vary along time, but we do not consider this concern in this work.

Table 1. Cost, emission, and standard deviations by technology. Source: Authors' own work, based on data collected from DeLlano et al. (2014, 2015) [14,77].

Technology and Abbreviation Used	Cost (€/MWh)	Cost Variance	Emission (kg-CO$_2$/MWh)	Emission Cost Variance
Nuclear (N)	30.04	8.07	-	-
Wind (W)	60.69	41.69	-	-
Offshore Wind (OW)	73.81	52.04	-	-
Hydro (H)	38.62	105.79	-	-
Small Hydro (SH)	42.95	12.92	-	-
PV	212.03	110.27	-	-
Biomass (B)	96.62	162.84	1.84	0.01
Coal (C)	52.23	31.51	734.09	4.77
Coal with CCS (C CCS)	78.44	46.27	101.00	0.66
Natural Gas (NG)	38.79	12.33	356.07	2.31
Natural Gas with CCS (NG CCS)	63.60	44.45	48.67	0.32
Oil (O)	93.17	155.83	546.46	3.55

A power-generation portfolio is a specific set of participation shares or weights of every technology in the model. For technology i, with $i = \{1, 2, \ldots, 12\}$, its participation share will be denoted by w_i. $w \in \mathbb{R}^{12 \times 1}$ represents the vector with twelve participation shares of a specific portfolio. These participation shares are the unknown variables of which the values are to be determined.

With $c \in \mathbb{R}^{12 \times 1}$, the vector containing the expected cost associated to every generation technology, the cost of a portfolio P can be calculated by Equation (1), where the supraindex t indicates the transposition operation.

$$c_P = w^t \times c \tag{1}$$

Now denote by $V \in \mathbb{R}^{12 \times 12}$ the variance-covariance matrix of the twelve technologies in the model. Thus, the portfolio risk, the standard deviation of the portfolio cost, will be as shown in Equation (2). Table 2 contains the cost variances-covariances used in the model.

$$\sigma_P = (w^t \times V \times w)^{\frac{1}{2}} \tag{2}$$

Table 2. Cost variance-covariance matrix. Source: Authors' own work, based on data collected from DeLlano et al. (2014, 2015) [14].

Technology	N	C	C CCS	NG	NG CCS	O	W	H	SH	OW	B	PV
N	8.07	3.84	5.07	3.54	4.26	15.32	−0.07	−0.42	−0.46	−0.10	−6.40	0.20
C	3.84	31.51	7.04	4.02	4.81	20.82	−0.21	0.03	0.03	−0.31	−14.1	-0.21
C CCS	5.07	7.04	46.27	5.43	6.60	27.16	−0.45	0.06	0.07	−0.68	−18.5	-0.46
NG	3.54	4.02	5.43	12.33	6.55	15.44	0.00	−0.08	−0.08	0.00	−3.16	0.05
CCS NG	4.26	4.81	6.60	6.55	44.45	18.33	0.00	−0.16	−0.17	0.00	−3.38	0.11
O	15.32	20.82	27.16	15.44	18.33	155.8	−4.02	−1.95	−2.11	−6.07	−86.4	-0.16
W	−0.07	−0.21	−0.45	0.00	0.00	−4.02	41.69	0.94	1.01	4.68	−0.31	0.09
H	−0.42	0.03	0.06	−0.08	−0.16	−1.95	0.94	105.8	3.64	1.41	−0.33	0.56
SH	−0.46	0.03	0.07	−0.08	−0.17	−2.11	1.01	3.64	12.92	1.53	−0.36	0.60
OW	−0.10	−0.31	−0.68	0.00	0.00	−6.07	4.68	1.41	1.53	52.04	−0.48	0.13
B	−6.40	−14.1	−18.5	−3.16	−3.38	−86.4	−0.31	−0.33	−0.36	−0.48	162.8	0.25
PV	0.20	−0.21	−0.46	0.05	0.11	−0.16	0.09	0.56	0.60	0.13	0.25	110.3

The problem of minimising the risk can be expressed in terms of a constrained quadratic optimization problem: $\min \sigma_P$, subject to a set of constraints that are described hereunder.

When applying MPT to power-generation planning, there are some technical constraints to keep in mind. First, the total sum of every participation share w_i must be equal to one. Also, every participation share w_i must be 0 or positive, and from the first constraint, it must also comply with $w_i \leq 1$.

The use of technologies for power generation is usually limited for the sake of the necessary power supply security, one of the main objectives of a country or territory power policy. Diversification of power-generation technologies leads inarguably to a more secure power supply. Moreover, another aim of a country or territory power policy is to preserve the environment and this can be reached by imposing stricter limits to the most pollutant technologies. Hence, our model has a set of technological and environmental constraints, imposing an upper limit on those weights w_i. These limits come from some diverse institutional forecasts (IEA, EU-IPTS, and the European Union Commission) and should be taken as reference limits as they can be adapted to specific country demands and as they are subject to changes over time. Table 3 details these limits.

Table 3. Limits by generation technology. Source: Authors' own work, based on data collected from DeLlano et al. (2015) [14].

Technology	Maximum Participation
Nuclear	29.80%
Wind	20.30%
Offshore Wind	2.00%
Hydro	10.80%
Small Hydro	1.50%
PV	5.5%
Biomass	8.50%
Coal with and without CCS	23.40%
Natural Gas with and without CCS	27.60%
Oil	0.80%
CCS technologies as a whole	18% of the non-CCS coal and natural gas, and oil participations

Just by using the aforementioned constraints—technical, technological, and environmental constraints—we are able to obtain a unique portfolio called the global minimum variance portfolio or

GMV portfolio. This portfolio shows the least risk level of every possible portfolio. As we want to obtain not only the GMV portfolio but also the set of efficient portfolios or efficient frontier, we must add an additional constraint to our model: the cost constraint. As a matter of fact, the GMV portfolio is the portfolio with the least risk but it is also an efficient portfolio with the highest cost. On the opposite end of the efficient frontier, we will find a portfolio with the lowest cost of all the efficient portfolios—but also with the highest risk of all the efficient portfolios. We can easily find this global minimum cost (GMC) portfolio by solving a linear programming problem of which the objective is to minimise the portfolio cost, $\min c_P = \min w^t \times c$, subject to the same constraints described above, except the cost constraint. This constraint is added to the quadratic model as $c_P = c^*$, with c^* as an objective cost for the portfolio. Iterating the quadratic model by changing this objective cost between the GMV cost and the GMC cost, we are able to draw the efficient frontier.

The model can then be expressed as in Equation (3).

$$\min \sigma_P = \min(w \times V \times w^t)^{\frac{1}{2}}, \text{ subject to :}$$

$$\begin{cases} w_i \geq 0, \ \forall i, \ i = \{1, 2, \ldots, 12\} \\ \sum_{i=1}^{12} w_i = 1 \\ \text{Technological and environmental constraints} \\ [c_P = w^t \times c = c^*] \end{cases} \tag{3}$$

Solving this model, we obtain the efficient frontier shown in Figure 1, where we also draw the extreme points of this frontier—the GMV portfolio and the GMC portfolio—along with their cost and risk values. In the graph, we also represent the cost-risk points corresponding to those technologies that fit into the graph's limits—nuclear, natural gas, and small hydro.

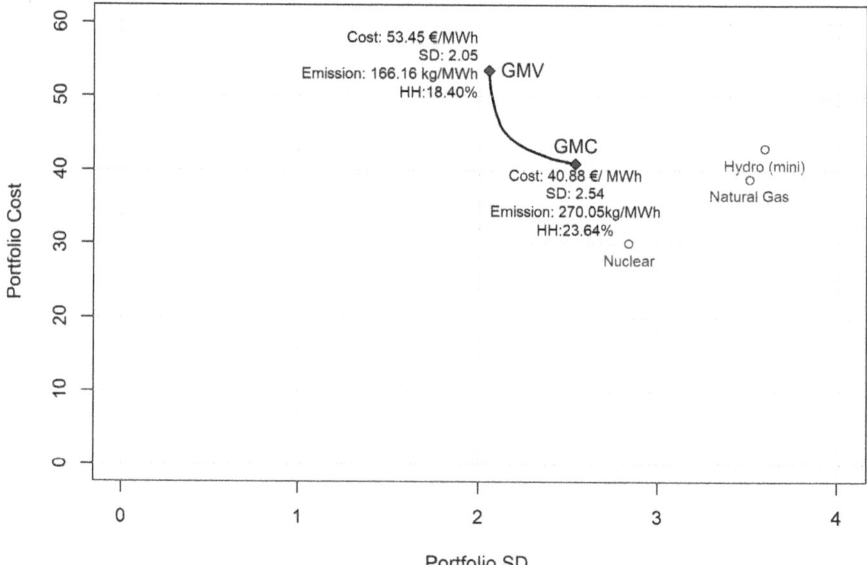

Figure 1. Instrumental model's efficient frontier.

Figure 2 represents the participation shares of the different technologies in the efficient frontier portfolios of the instrumental model. As we can see, the GMV portfolio, the one on the left side of the figure, shows a higher diversification—its Herfindahl–Hirschman index is 18.40%, less than the GMC portfolio with a Herfindahl–Hirschman index of 23.64%, which is considered good for energy security.

Weights in the efficient frontier from GMV to GMC portfolios

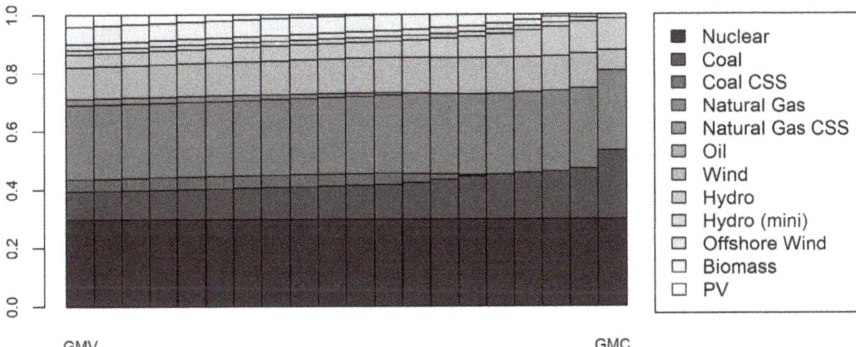

Figure 2. Technology participation in the instrumental mode.

Moreover, the nuclear, the small hydro, as well as the offshore wind technologies participate in the GMV portfolio at their maxima. On the other hand, in the GMC portfolio, the nuclear, the coal, the natural gas, the hydro, the small hydro, and the offshore wind technologies play a part at their maxima. Therefore, nuclear, small hydro, and offshore wind are regarded as highly efficient technologies in terms of cost and risk by the model.

The next step will be to classify the different generation technologies into two different subsets: one for the pollutant technologies—coal, coal with CCS, natural gas, natural gas with CCS, oil, and biomass—and another one for the non-pollutant technologies—nuclear, wind, offshore wind, hydro, small hydro, and PV. For the first subset, we are going to set out a model quite similar to the instrumental one just exposed. For the second subset, we will consider an emission-risk model instead of a cost-risk model.

2.3.2. The Non-Pollutant Technology Efficient Frontier

Using the same data shown in Tables 1 and 2 but considering only those non-pollutant technologies, we will adapt the technological and environmental constraints of the instrumental model to obtain the constraints pertinent to this non-pollutant technologies model. To adapt the technological and environmental constraints of the instrumental model to the non-pollutant technologies, we decided to work on the basis of the non-pollutant technology's participation shares in the efficient frontier calculated in the previous section. Briefly, we raised the participation share of every non-pollutant technology relative to the total non-pollutant technologies participation shares in every efficient portfolio and took the maximum by technology, resulting in the limits presented in Table 4, where the column "Maximum Weight" refers to the maximum weight reached in the efficient frontier of the previous model and the column "Maximum Participation" refers to this maximum weight considering only the non-pollutant technologies.

Table 4. Non-pollutant technological and environmental limits. Source: Authors' own work, based on data collected from DeLlano et al. (2015) [14].

Technology	Maximum Weight	Maximum Participation
Nuclear	29.80%	60.82%
Wind	12.66%	23.86%
Offshore Wind	2.00%	3.92%
Hydro	10.80%	22.04%
Small Hydro	1.50%	3.06%
PV	4.16%	7.86%

Thus, the non-pollutant model is presented in Equation (4); keep in mind that the cost restriction is used to calculate the efficient frontier as described in the previous section.

$$\text{min}\sigma_P = \min(w \times V \times w^t)^{\frac{1}{2}}, \text{ subject to :}$$

$$\begin{cases} w_i \geq 0,\ \forall i,\ i = \{1,\, 2,\, \ldots,\, 6\} \\ \sum_{i=1}^{6} w_i = 1 \\ w_{\text{Nuclear}} \leq 60.82\% \\ w_{\text{Wind}} \leq 23.86\% \\ w_{\text{Offshore Wind}} \leq 3.92\% \\ w_{\text{Hydro}} \leq 22.04\% \\ w_{\text{Small Hydro}} \leq 3.06\% \\ w_{\text{PV}} \leq 7.86\% \\ [c_P = w^t \times c = c^*] \end{cases} \qquad (4)$$

Solving this model, we obtain the efficient frontier shown in Figure 3. Note that the dashed grey line shown is the instrumental model efficient frontier from the previous section. Comparing both efficient frontiers, the one from the instrumental model and the one from the non-pollutant technology model, we can observe that the instrumental model shows higher costs but lower risks—as its efficient frontier is displaced upward and toward the left.

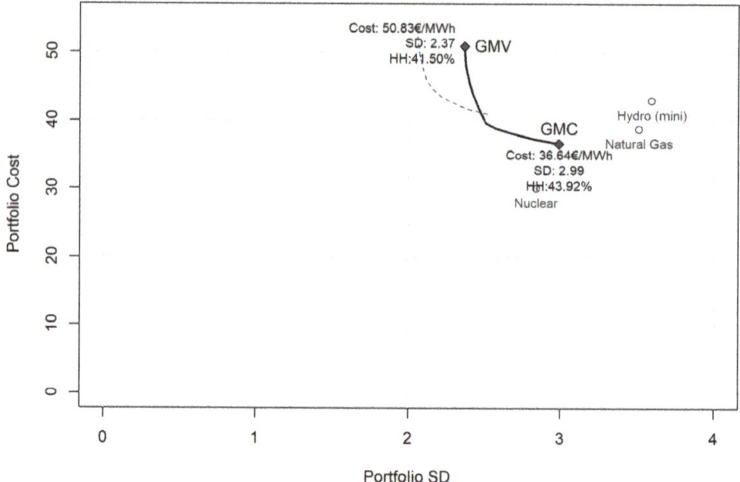

Figure 3. Non-pollutant technology model's efficient frontier.

Analysing the weights in the GMV and GMC portfolios, we see that nuclear and small hydro participate again at their maxima in both portfolios. In particular, in the GMV portfolio, the offshore wind also participates at its maximum, while in the GMC portfolio, it is the hydro technology that also enters at its maximum.

Not surprisingly, the Herfindahl–Hirschman index is worse than in the instrumental model, the non-pollutant technology model has fewer technologies, and again, the GMV portfolio is more diversified than the GMC portfolio.

2.3.3. The Emission-Risk Pollutant Technology Efficient Frontier

We will replace our cost-risk orientation with an emission-risk orientation in this second stage of the model presented in this work. As stated, we have no emission data, apart from the emission average shown in Table 1, and hence, we use the CO_2 cost standard deviation as a proxy for the emission standard deviation. According to this information, we simulated 100,000 normal distributed values to calculate the variance-covariance matrix shown in Table 5.

Table 5. Emission variance-covariance matrix.

Technology	Coal	Coal with CCS	Natural Gas	Natural Gas with CCS	Oil	Biomass
Coal	22.846215	−0.014355	−0.020952	−0.003395	−0.004446	0.000102
Coal with CCS	−0.014355	0.436912	0.004161	0.000500	0.007233	0.000058
Natural Gas	−0.020952	0.004161	5.298606	−0.001473	0.002642	−0.000033
Natural Gas with CCS	−0.003395	0.000500	−0.001473	0.102033	0.002162	−0.000014
Oil	−0.004446	0.007233	0.002642	0.002162	12.594664	0.000114
Biomass	0.000102	0.000058	−0.000033	−0.000014	0.000114	0.000100

The pollutant technology emission-risk model presented in this section is highly similar to the instrumental model except for the following two aspects. Firstly, we are not using costs in the current model but emission factors for the six pollutant technologies—coal, coal with CCS, natural gas, natural gas with CCS, oil, and biomass. Secondly, we will show five different adaptations of the current model, four of them without constraints other than the model technical constraints. In the other model adaptation, we include technological constraints for the pollutant technologies. The limits of these constraints were built in a quite similar manner as that for the non-pollutant technology model presented in Section 2.3.2., i.e., raising every pollutant technology participation in the instrumental model efficient frontier portfolios relative to the total pollutant technologies in those portfolios and getting the maximum participation as the limit. A problem arises as the oil technology is not participating in any of the calculated portfolios. To fix it, its technological and environmental limits in the instrumental model are relatively raised to the limits set for the pollutant technologies.

2.3.4. Model Adaptation with All the Non-Pollutant Technologies

For the six pollutant technologies, we solved the model presented in Equation (5), in which $e_P \in \mathbb{R}$ is the emission factor of the portfolio and $e \in \mathbb{R}^{6 \times 1}$ is the emission vector of which the elements can be found in Table 1:

$$\min \sigma_P = \min(w \times V \times w^t)^{\frac{1}{2}}, \text{ subject to :}$$
$$\begin{cases} w_i \geq 0, \ \forall i, \ i = \{1, 2, \ldots, 6\} \\ \sum_{i=1}^{6} w_i = 1 \\ [e_P = w^t \times e = e^*] \end{cases} \tag{5}$$

In this model we substituted the cost constraint for an emission constraint as we are working with emission-risk pairs instead of cost-risk pairs.

The results of this model are trivial because, in the GMV portfolio, 99.88% of the power generation is assigned to biomass and, in the global minimum emission portfolio (GME), biomass captures 100% of the power generation. The efficient frontier is hence insignificant and practically indistinguishable from a portfolio with 100% biomass generation. These results were expected as biomass shows a very low level of CO_2 emission as compared to the rest of pollutant technologies and a negligible level of risk. In fact, this result is completely in line with the optimization features of the model.

A single-technology generation portfolio, or a generation portfolio in which one single technology is responsible for such a big part of the power generation, is quite far from being an acceptable solution from the point of view of energy planning. Next, we develop some model adaptations to deal with this circumstance.

2.3.5. Model Adaptations without CCS Technologies and without Biomass Technology

The first adaptation is similar to the previous one, but CCS technologies, both coal and natural gas, are removed to prevent the possibility of these technologies not being able to reach a feasible commercial availability. The results are therefore similar: biomass hoards 99.99% of the generation in the GMV portfolio and 100% of the generation in the GME portfolio because of the reasons exposed above.

In the second model adaptation, biomass technology is removed from the model. The results offer a bit more information than in the previous models. Figure 4 shows the weights of the five considered technologies from the GMV portfolio—first column on the left—to the GME portfolio—last column. The GMV portfolio allows entry of every technology to the generation mix although natural gas with CCS takes the lion's share—the Herfindahl–Hirschman index is 65.96% for this portfolio. As we move from the GMV portfolio to the GME portfolio, it is clear that natural gas with CCS is increasing its participation share until it reaches 100% in the GME portfolio.

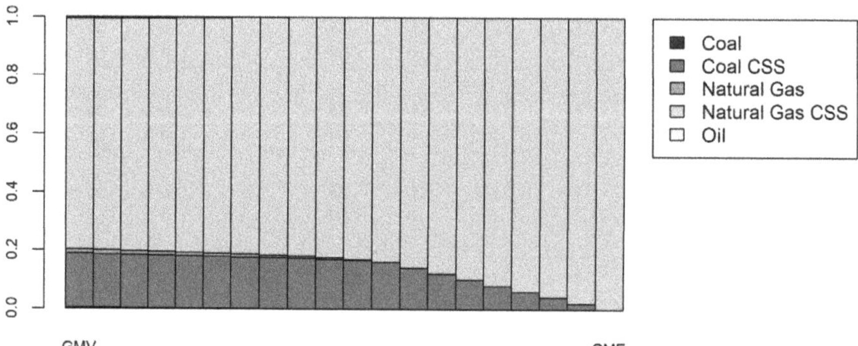

Figure 4. Non-pollutant technology model's efficient frontier.

The third model adaptation shows what happens if neither CCS technologies nor biomass are available to generate power. Again, natural gas, in this case without CCS, is the dominant generation technology. Surprisingly, oil takes the second place. This is due to the correlations between oil and the other two technologies in this model adaptation. Figure 5 shows the weights of the considered technologies in the efficient frontier portfolios—from the GMV portfolio on the left to the GME portfolio on the right.

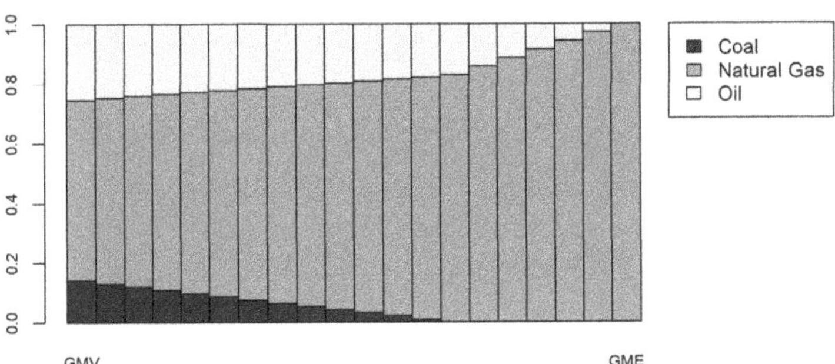

Figure 5. Model adaptation without both carbon capture and storage (CCS) and biomass technologies.

2.3.6. Model Adaptation with Technological Constraints

In this model adaptation, the problem to solve will be the one presented in Equation (6).

$$\min \sigma_P = \min(w \times V \times w^t)^{\frac{1}{2}}, \text{ subject to :}$$

$$\begin{cases} w_i \geq 0, \ \forall_i, \ i = \{1, 2, \ldots, 6\} \\ \sum_{i=1}^{6} w_i = 1 \\ w_{\text{Coal}} + w_{\text{Coal with with CCS}} \leq 54.84\% \\ w_{\text{Natural gas}} + w_{\text{Natural gas with CCS}} \leq 66.40\% \\ w_{\text{Oil}} \leq 1.33\% \\ w_{\text{Biomass}} \leq 12.56\% \\ [e_P = w^t \times e = e^*] \end{cases} \tag{6}$$

As stated, the limits were taken from the instrumental model efficient frontier participation shares of the pollutant technologies, except for the oil technology limit that was taken from the instrumental model technological limits of the pollutant technologies.

Solving this model adaptation, the results shown in Figure 6 are achieved. As expected, in light of the precedent results, biomass is participating at its maximum in every efficient frontier portfolio. When they participate in the less risky portfolios, coal, natural gas, and oil have participation shares around 1%. In the GME portfolio, natural gas with CCS participates at the maximum set for natural gas with and without CCS. The little variation in participation shares due to the imposed constraints will give us a short efficient frontier.

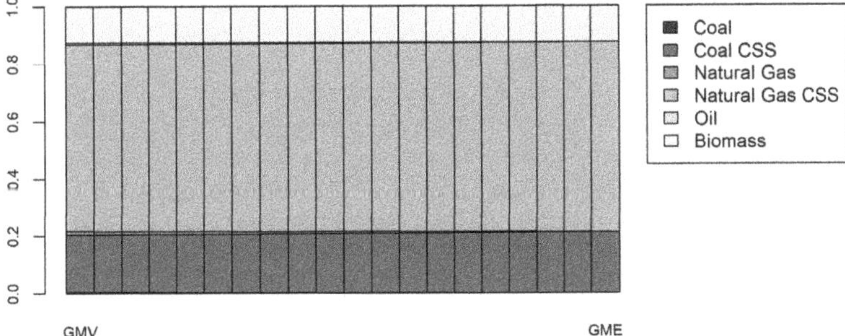

Figure 6. Model adaptation with technological constraints.

3. Results

In this section, we present our main results related to cross-drawing the instrumental model and the pollutant-technology model. Additionally, we will discuss how this model can help policy makers make their decisions.

3.1. Cross-Drawing the Cost-Risk and Emission-Risk Models and Selecting an Adequate Combination of Non-Pollutant and Pollutant Technologies

So far, we have one instrumental model that includes all the technologies and constraints, a non-pollutant efficient frontier that shows higher risk but lower cost than the instrumental efficient frontier, and a set of several adaptations of a model with pollutant-technologies. Figure 7 represents some of the efficient frontiers calculated in an emission-risk coordinate axis. Specifically, we depict the instrumental model numbered with 0 and with a dot-dash line, the model adaptations without biomass numbered as 2.c, those without biomass and CSS technologies numbered as 2.d, and those with all the pollutant technologies and with technological constraints numbered as 2.e. It is important to note that the first two adaptations were practically 100% biomass participated, and for this reason, we are not showing them in the graph—they would be located practically where the biomass technology is drawn.

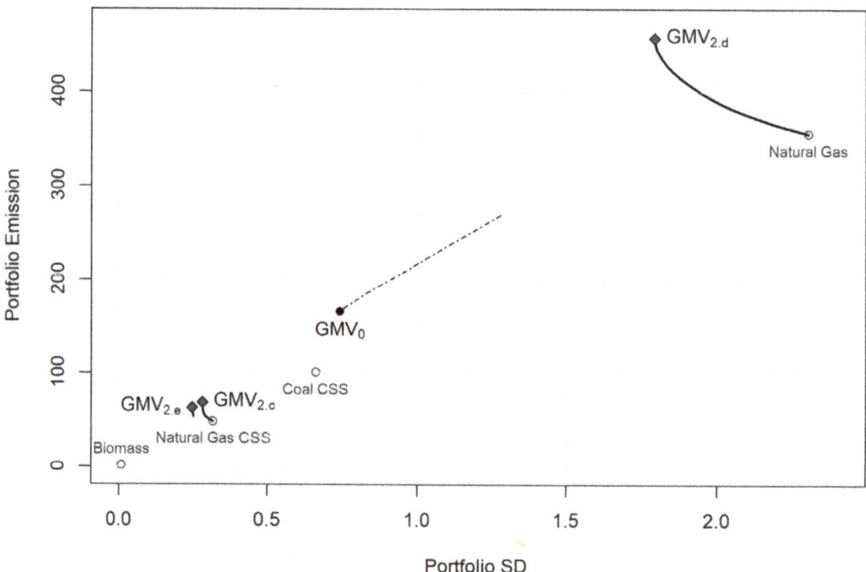

Figure 7. Efficient frontiers in an emission-risk plane.

Regarding the pollutant models, the traditional pollutant technologies—coal, natural gas, and oil—show higher levels of emission and risk; model 2.d efficient frontier appears on the top right side of the figure. If CCS technologies are included, both the emission and the risk levels are drastically reduced; see model 2.c in the figure. In fact, models 2.a and 2.b would be represented over the biomass point in Figure 6. Moreover, the technological constraints are able to lower even more the risk, keeping a similar level of emission—model 2.e.

By representing in the same emission-risk plane our instrumental model, model 0, it is worth comparing it with the pollutant models—the non-pollutant model would be drawn on the coordinate origin. The instrumental model shows a higher level of emission and risk, in terms of emission, than those models allowing biomass and CCS technologies because coal and natural gas participate largely in it, as shown in Figure 2. When approaching the GMC portfolio, these technologies reach their technological limit and, actually, they participate at their maxima in the GMC portfolio.

The efficient frontier of our models is drawn in a cost-risk coordinate axis in Figure 8. Both the instrumental model, model 0, and the pollutant models, models 2.e and 2.d, present smaller levels of risk with similar or lower levels of cost. As stated, pollutant models with biomass, models 2.a and 2.b, would be drawn on the point corresponding to biomass technology that is far out of the graph's limits with a cost variance of 162.84 (standard deviation: 12.76 €/MWh) and with a cost of 96.62 €/MWh.

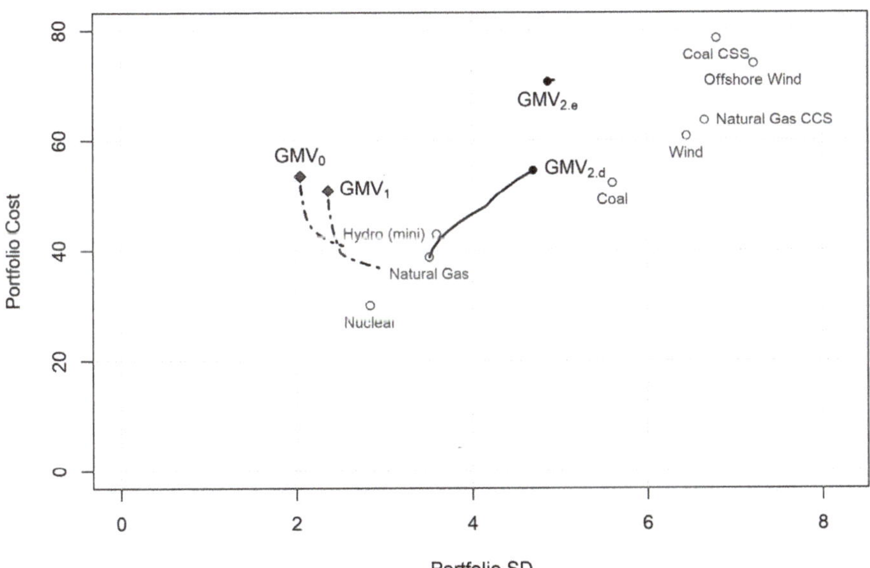

Figure 8. Efficient frontiers in a cost-risk plane.

3.2. The CML-Analogous Area

So far, we have a pollutant-technology efficient frontier from an emission-risk perspective and a point of the emission-risk coordinate axis origin representing all the non-pollutant efficient portfolios. A policy maker could compile a portfolio from the pollutant-technology efficient frontier with the point in the origin to determine a power-generation portfolio with the whole set of technologies. Therefore, it is possible to set the best portfolio given a desired emission factor or a risk limit.

The limits of the pollutant-technology efficient frontier are the GMV and the GME portfolios. The efficient frontier itself connects them together. Combinations of either the GMV or the GME portfolio with any of the non-pollutant efficient portfolios in the origin will fall inside an area delimited by these three portfolios: the GMV portfolio, the GME portfolio, and the non-pollutant efficient portfolio chosen. In Figure 9, this area is the shaded area below and to the left of the pollutant-technology efficient frontier.

Being under the efficient frontier reflects that any point inside that area shows a lower emission factor than the point on the frontier with the same level of risk. This was expected as we are combining a pollutant portfolio with a non-pollutant one. On the other hand, the fact of being to the left of the pollutant efficient frontier indicates that the risk is lower for any emission factor considered. A portfolio inside the CML-analogous area (CML-A) is then more efficient than a portfolio in the efficient frontier with the same emission factor or level of risk.

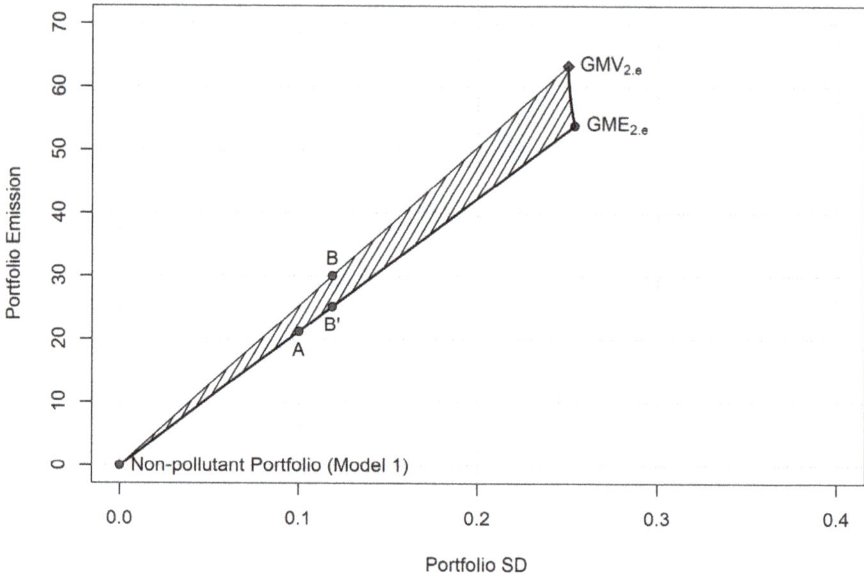

Figure 9. A Capital Market Line (CML)-Analogous analysis.

Focusing on the CML-A, the problem is to determine the best portfolio for a given emission factor or for a given level of risk. It is easy to conclude that the answer must be found on the CML-A borders. Indeed, when determining the best portfolio in the CML-A for a given level of risk, the solution must be that one located on the segment joining the coordinate axis origin and the GME portfolio that shows that level of risk. Likewise, if we want to determine the best portfolio in the CML-A for a given emission factor, we must find it on the intersection of the segment joining the GMV portfolio with the coordinate axis origin and the line representing the desired emission factor. In the next section, we present a brief example of these ideas.

4. Discussion

In this section, there is a brief explanation of how a policy maker could employ this model to design power-generation portfolios. In Figure 9, we draw one of our pollutant efficient frontiers, specifically model 2.e, with all the pollutant technologies and with technological constraints in an emission-risk coordinate axis. In this graph, the non-pollutant model portfolios will be all of them on the coordinate origin; they have no emission and, consequently, no emission risk.

A policy maker can choose any combination: any linear combination between any portfolio on the pollutant efficient frontier and on the non-pollutant efficient frontier which, from an emission point of view, lays on the coordinate origin. In Figure 9, the shaded area delimited by the coordinate origin and the pollutant efficient frontier of model 2.e represents these combinations. We can see that the area allows the policy maker to lower the risk for a given level of emission or the emission for a given level of risk. For instance, given the emission and risk values for GMV and GME portfolios of model 2.e shown in Table 6, it is easy to calculate the portfolio proportions needed to reach an emission risk of 0.10 kg/MWh. Obviously, the policy maker will prefer the lowest emission possible for that level of risk and so they will choose the pollutant GME portfolio for the combination, resulting in point A in Figure 9. Point A can be reached by combining the pollutant GME portfolio and the non-pollutant portfolio in a proportion of 39.39–60.61%. Also, the policy maker could want to set the emission of the resultant portfolio, say, at 30 kg/MWh. In this case, they would like to reach a minimum risk combination for that level of emission. For this reason, they will want to combine the pollutant GMV portfolio with the non-pollutant portfolio, resulting in point B, which corresponds to a 47.44–52.56% proportion of pollutant GMV and non-pollutant portfolios.

Table 6. Emission and risk of the global minimum variance (GMV) and global minimum emission (GME) portfolios of model 2.e.

Portfolio	Emission (kg-CO$_2$/MWh)	Risk (kg-CO$_2$/MWh)
GMV$_{2.e}$	63.24	0.2502
GME$_{2.e}$	53.79	0.2539

In the last example, it is easy to see that the combination B′ with a 46.75% GME portfolio and a 53.25% non-pollutant portfolio has the same risk value as combination B but with a lower emission of 25.15 kg/MWh. It does not seem reasonable to lose the opportunity to lower emissions without increasing the risk. This is why the lower segment of the shaded area is more efficient in the sense used in this work than the rest of the points in the area when the aim is to adapt the generation portfolio to a predetermined risk. This insight is similar to the financial CML, but in our case and due to the convexity of the efficient frontier, instead of having a tangency or market portfolio, we propose to use the corresponding GME portfolio instead.

5. Conclusions

Throughout the present work, we proved that it is possible to separate the generation technologies into two different sets and to proceed to a double optimization of the generation mix. When we compare the non-pollutant-technology efficient frontier with the efficient frontier of the instrumental model, we are able to generate at a lower cost but at a higher risk using only non-pollutant technologies (nuclear, wind, offshore wind, hydro, small hydro, and PV).

When analysing the sharing weights in the non-pollutant efficient frontier, nuclear energy, defending its position as a base-load generation technology, and small hydro participate at their maxima in both the minimum-risk GMV and the minimum-cost GMC portfolios:

- Nuclear and small hydro are preferential technologies that act as if it intends to obtain the minimum cost or to get the minimum risk of the portfolio.
- In a complementary way, offshore wind technology participates at its maximum share if the minimum risk is searched, while large hydro technology is the third technology to enter its maximum in the minimum cost portfolio.

Replacing the cost-risk perspective with emission-risk perspective pollutant technologies allows to highlight the important role of biomass and CCS technologies in an efficient portfolio. Their commercial development is crucial in order to achieve low-carbon emission portfolios.

Oil generation is not included in the power-generation mix in the instrumental model, highlighting its excessive cost and risk. In the emission-risk models, it is only considered when we take out the biomass or when we set upper limits to the participation shares of the technologies. These limits cause the preferred technologies to participate at their maxima in almost every efficient portfolio.

Solar PV generation takes part only in the efficient portfolios close to the GMV portfolio. Its participation is needed in order to achieve a highly diversified and lower risk portfolio.

The cross-drawing approach proposed between the pollutant and non-pollutant efficient frontiers calculated in both cost-risk and emission-risk coordinate axes leads to relevant conclusions:

- A pollutant-only generation mix shows a higher cost than a complete generation technology portfolio and even in relation to the non-pollutant-only efficient frontier.
- A highly diversified portfolio makes it possible to achieve the lowest risk (instrumental model).

- Renewable energy sources are needed to reduce portfolio cost and risk.
- Pollutant-generation-efficient frontiers show a higher risk mainly because of the fuel cost risk.

Finally, drawing an analogy with the CML from CAPM, we presented the CML-A area that could helpful a policy maker design the long-term generation mix in a decarbonisation scenario.

Author Contributions: The Contributor Roles Taxonomy (CRediT) of this work is as follows: conceptualization, P.M.-F., F.d.-P. and I.S.; methodology: P.M.-F., F.d.-P. and A.C.-S.; software, P.M.-F.; validation, F.d.-P.; formal analysis, P.M.-F., F.d.-P. and A.C.-S.; investigation: F.d.-P.; resources, F.d.-P. and I.S.; data curation, P.M.-F. and F.d.-P.; writing – original draft preparation, P.M.-F. and F.d.-P.; writing – review and editing, P.M.-F.; visualization, P.M.-F. and F.d.-P.; supervision, A.C.-S. and I.S.; project administration, P.M.-F. and F.d.-P.; funding acquisition: A.C.-S.

Funding: This research received no external funding.

Conflicts of Interest: The authors declare no conflict of interest.

References

1. Cormio, C.; Dicorato, M.; Minoia, A.; Trovato, M. A regional energy planning methodology including renewable energy sources and environmental constraints. *Renew. Sustain. Energy Rev.* **2003**, *7*, 99–130. [CrossRef]
2. Huang, Y.-H.; Wu, J.-H.; Hsu, Y.-J. Two-stage stochastic programming model for the regional-scale electricity planning under demand uncertainty. *Energy* **2016**, *116*, 1145–1157. [CrossRef]
3. Liu, Y.; He, L.; Shen, J. Optimization-based provincial hybrid renewable and non-renewable energy planning—A case study of Shanxi, China. *Energy* **2017**, *128*, 839–856. [CrossRef]
4. Nie, S.; Li, Y.P.; Liu, J.; Huang, C.Z. Risk management of energy system for identifying optimal power mix with financial-cost minimization and environmental-impact mitigation under uncertainty. *Energy Econ.* **2017**, *61*, 313–329. [CrossRef]
5. Kim, S.; Koo, J.; Lee, C.J.; Yoon, E.S. Optimization of Korean energy planning for sustainability considering uncertainties in learning rates and external factors. *Energy* **2012**, *44*, 126–134. [CrossRef]
6. Codina Gironès, V.; Moret, S.; Maréchal, F.; Favrat, D. Strategic energy planning for large-scale energy systems: A modelling framework to aid decision-making. *Energy* **2015**, *90*, 173–186. [CrossRef]
7. Sáfián, F. Modelling the Hungarian energy system—The first step towards sustainable energy planning. *Energy* **2014**, *69*, 58–66. [CrossRef]
8. Gómez, A.; Dopazo, C.; Fueyo, N. The "cost of not doing" energy planning: The Spanish energy bubble. *Energy* **2016**, *101*, 434–446. [CrossRef]
9. Allan, G.; Eromenko, I.; McGregor, P.; Swales, K. The regional electricity generation mix in Scotland: A portfolio selection approach incorporating marine technologies. *Energy Policy* **2011**, *39*, 6–22. [CrossRef]
10. Arnesano, M.; Carlucci, A.P.; Laforgia, D. Extension of portfolio theory application to energy planning problem—The Italian case. *Energy* **2012**, *39*, 112–124. [CrossRef]
11. Awerbuch, S. Portfolio-Based Electricity Generation Planning: Implications for Renewables and Energy Security. *Mitig. Adapt. Strateg. Glob. Chang.* **2004**, *11*, 693–710. [CrossRef]
12. Awerbuch, S.; Jansen, J.C.; Beurskens, L. The Role of Wind Generation in Enhancing Scotland's Energy Diversity and Security: A Mean-Variance Portfolio Optimization of Scotland's Generating Mix. In *Analytical Methods for Energy Diversity & Security*; Elsevier: Amsterdam, The Netherlands, 2008; pp. 139–150.
13. Bhattacharya, A.; Kojima, S. Power sector investment risk and renewable energy: A Japanese case study using portfolio risk optimization method. *Energy Policy* **2012**, *40*, 69–80. [CrossRef]
14. Dellano-Paz, F.; Calvo-Silvosa, A.; Iglesias Antelo, S.; Soares, I. The European low-carbon mix for 2030: The role of renewable energy sources in an environmentally and socially efficient approach. *Renew. Sustain. Energy Rev.* **2015**, *48*, 49–61. [CrossRef]
15. Delarue, E.; De Jonghe, C.; Belmans, R.; D'haeseleer, W. Applying portfolio theory to the electricity sector: Energy versus power. *Energy Econ.* **2011**, *33*, 12–23. [CrossRef]
16. Dellano-Paz, F.; Calvo-Silvosa, A.; Antelo, S.I.; Soares, I. Energy planning and modern portfolio theory: A review. *Renew. Sustain. Energy Rev.* **2017**, *77*, 636–651. [CrossRef]
17. Kumar, D.; Mohanta, D.K.; Reddy, M.J.B. Intelligent optimization of renewable resource mixes incorporating the effect of fuel risk, fuel cost and CO_2 emission. *Front. Energy* **2015**, *9*, 91–105. [CrossRef]

18. Gao, C.; Sun, M.; Shen, B.; Li, R.; Tian, L. Optimization of China's energy structure based on portfolio theory. *Energy* **2014**, *77*, 890–897. [CrossRef]
19. Gökgöz, F.; Atmaca, M.E. Financial optimization in the Turkish electricity market: Markowitz's mean-variance approach. *Renew. Sustain. Energy Rev.* **2012**, *16*, 357–368. [CrossRef]
20. Lucheroni, C.; Mari, C. CO_2 volatility impact on energy portfolio choice: A fully stochastic LCOE theory analysis. *Appl. Energy* **2017**, *190*, 278–290. [CrossRef]
21. Roques, F.; Hiroux, C.; Saguan, M. Optimal wind power deployment in Europe—A portfolio approach. *Energy Policy* **2010**, *38*, 3245–3256. [CrossRef]
22. Zhu, L.; Fan, Y. Optimization of China's generating portfolio and policy implications based on portfolio theory. *Energy* **2010**, *35*, 1391–1402. [CrossRef]
23. Chen, Y.; He, L.; Guan, Y.; Lu, H.; Li, J. Life cycle assessment of greenhouse gas emissions and water-energy optimization for shale gas supply chain planning based on multi-level approach: Case study in Barnett, Marcellus, Fayetteville, and Haynesville shales. *Energy Convers. Manag.* **2017**, *134*, 382–398. [CrossRef]
24. Shaban Boloukat, M.H.; Akbari Foroud, A. Stochastic-based resource expansion planning for a grid-connected microgrid using interval linear programming. *Energy* **2016**, *113*, 776–787. [CrossRef]
25. Falsafi, H.; Zakariazadeh, A.; Jadid, S. The role of demand response in single and multi-objective wind-thermal generation scheduling: A stochastic programming. *Energy* **2014**, *64*, 853–867. [CrossRef]
26. Koltsaklis, N.E.; Liu, P.; Georgiadis, M.C. An integrated stochastic multi-regional long-term energy planning model incorporating autonomous power systems and demand response. *Energy* **2015**, *82*, 865–888. [CrossRef]
27. Tajeddini, M.A.; Rahimi-Kian, A.; Soroudi, A. Risk averse optimal operation of a virtual power plant using two stage stochastic programming. *Energy* **2014**, *73*, 958–967. [CrossRef]
28. Stempien, J.P.; Chan, S.H. Addressing energy trilemma via the modified Markowitz Mean-Variance Portfolio Optimization theory. *Appl. Energy* **2017**, *202*, 228–237. [CrossRef]
29. Jansen, J.C.; Beurskens, L.W.M.; Van Tilburg, X. *Application of Portfolio Analysis to the Dutch Generating Mix; Reference case and two renewables cases, Year 2030, SE and GE scenario* (No. ECN-C-05-100); Energy research Centre of the Netherlands ECN: Sint Maartensvlotbrug, The Netherlands, 2006.
30. Dellano-Paz, F.; Martínez Fernandez, P.; Soares, I. Addressing 2030 EU policy framework for energy and climate: Cost, risk and energy security issues. *Energy* **2016**, *115*, 1347–1360. [CrossRef]
31. Hickey, E.A.; Lon Carlson, J.; Loomis, D. Issues in the determination of the optimal portfolio of electricity supply options. *Energy Policy* **2010**, *38*, 2198–2207. [CrossRef]
32. Jano-Ito, M.A.; Crawford-Brown, D. Investment decisions considering economic, environmental and social factors: An actors' perspective for the electricity sector of Mexico. *Energy* **2017**, *121*, 92–106. [CrossRef]
33. Chuang, M.C.; Ma, H.W. Energy security and improvements in the function of diversity indices—Taiwan energy supply structure case study. *Renew. Sustain. Energy Rev.* **2013**, *24*, 9–20. [CrossRef]
34. Böhringer, C.; Bortolamedi, M. Sense and no(n)-sense of energy security indicators. *Ecol. Econ.* **2015**, *119*, 359–371. [CrossRef]
35. Muñoz, B.; García-Verdugo, J.; San-Martín, E. Quantifying the geopolitical dimension of energy risks: A tool for energy modelling and planning. *Energy* **2015**, *82*, 479–500. [CrossRef]
36. Apergis, N.; Payne, J.E.; Menyah, K.; Wolde-Rufael, Y. On the causal dynamics between emissions, nuclear energy, renewable energy, and economic growth. *Ecol. Econ.* **2010**, *69*, 2255–2260. [CrossRef]
37. Johansson, B. Security aspects of future renewable energy systems–A short overview. *Energy* **2013**, *61*, 598–605. [CrossRef]
38. Labandeira, X. *Sistema Energético Y Cambio Climático: Prospectiva Tecnológica y Regulatoria (No. 02)*; Working Papers 2012; Economics for Energy (eforenergy.org): Vigo, Spain, 2012.
39. Labandeira, X.; Linares, P.; Würzburg, K. *Energías Renovables Y Cambio Climático (No. 06)*; Working Papers 2012; Economics for Energy (eforenergy.org): Vigo, Spain.
40. Akter, S.; Kompas, T.; Ward, M.B. Application of portfolio theory to asset-based biosecurity decision analysis. *Ecol. Econ.* **2015**, *117*, 73–85. [CrossRef]
41. Tolmasquim, M.T.; Seroa da Motta, R.; La Rovere, E.L.; Barata MM de, L.; Monteiro, A.G. Environmental valuation for long-term strategic planning—The case of the Brazilian power sector. *Ecol. Econ.* **2001**, *37*, 39–51. [CrossRef]
42. Urhammer, E. Celestial bodies and satellites—Energy issues, models, and imaginaries in Denmark since 1973. *Ecol. Econ.* **2017**, *131*, 425–433. [CrossRef]

43. Bennink, D.; Rooijers, F.; Croezen, H.; de Jong, F.; Markowska, A. *VME Energy Transition Strategy: External Costs and Benefits of Electricity Generation, Transition*; CE Delft: Delft, The Netherlands, 2010.

44. Eyre, N. External costs. What do they mean for energy policy? *Energy Policy* **1997**, *25*, 85–95. [CrossRef]

45. Söderholm, P.; Sundqvist, T. Pricing environmental externalities in the power sector: Ethical limits and implications for social choice. *Ecol. Econ.* **2003**, *46*, 333–350. [CrossRef]

46. Awerbuch, S.; Yang, S. *Efficient Electricity Generating Portfolios for Europe: Maximising Energy Security and Climate Change Mitigation*; EIB Papers; EIB: Kirchberg, Luxembourg, 2007.

47. Cucchiella, F.; D'Adamo, I.; Gastaldi, M. Optimizing plant size in the planning of renewable energy portfolios. *Lett. Spat. Resour. Sci.* **2016**, *9*, 169–187. [CrossRef]

48. Lynch, M.Á.; Shortt, A.; Tol, R.S.J.; O'Malley, M.J. Risk–return incentives in liberalised electricity markets. *Energy Econ.* **2013**, *40*, 598–608. [CrossRef]

49. Antimiani, A.; Costantini, V.; Kuik, O.; Paglialunga, E. Mitigation of adverse effects on competitiveness and leakage of unilateral EU climate policy: An assessment of policy instruments. *Ecol. Econ.* **2016**, *128*, 246–259. [CrossRef]

50. Monjon, S.; Quirion, P. Addressing leakage in the EU ETS: Border adjustment or output-based allocation? *Ecol. Econ.* **2011**, *70*, 1957–1971. [CrossRef]

51. Oberndorfer, U. EU Emission Allowances and the stock market: Evidence from the electricity industry. *Ecol. Econ.* **2009**, *68*, 1116–1126. [CrossRef]

52. Rogge, K.S.; Schneider, M.; Hoffmann, V.H. The innovation impact of the EU Emission Trading System—Findings of company case studies in the German power sector. *Ecol. Econ.* **2011**, *70*, 513–523. [CrossRef]

53. Dincer, I. Renewable energy and sustainable development: A crucial review. *Renew. Sustain. Energy Rev.* **2000**, *4*, 157–175. [CrossRef]

54. Escribano Francés, G.; Marín-Quemada, J.M.; San Martín González, E. RES and risk: Renewable energy's contribution to energy security. A portfolio-based approach. *Renew. Sustain. Energy Rev.* **2013**, *26*, 549–559. [CrossRef]

55. Panwar, N.L.; Kaushik, S.C.; Kothari, S. Role of renewable energy sources in environmental protection: A review. *Renew. Sustain. Energy Rev.* **2011**, *15*, 1513–1524. [CrossRef]

56. Martinez-Fernandez, P.; deLlano-Paz, F.; Calvo-Silvosa, A.; Soares, I. Pollutant versus non-pollutant generation technologies: A CML-analogous analysis. *Environ. Dev. Sustain.* **2018**, *20* (Suppl. 1), 199–212. [CrossRef]

57. Jaffe, A.B.; Newell, R.G.; Stavins, R.N. A tale of two market failures: Technology and environmental policy. *Ecol. Econ.* **2005**, *54*, 164–174. [CrossRef]

58. Markowitz, H. Portfolio Selection. *J. Financ.* **1952**, *7*, 77–91. [CrossRef]

59. Sharpe, W.F. A Simplified Model for Portfolio Analysis. *Manage. Sci.* **1963**, *9*, 277–293. [CrossRef]

60. Sharpe, W.F. Capital Asset Prices: A Theory of Market Equilibrium under Conditions of Risk. *J. Financ.* **1964**, *19*, 425. [CrossRef]

61. Min Liu Wu, F.F.; Yixin, N.I. A survey on risk management in electricity markets. In Proceedings of the 2006 IEEE Power Engineering Society General Meeting, Montreal, QC, Canada, 18–22 June 2006; IEEE: Piscataway, NJ, USA, 2006; p. 6. [CrossRef]

62. Westner, G.; Madlener, R. The benefit of regional diversification of cogeneration investments in Europe: A mean-variance portfolio analysis. *Energy Policy* **2010**, *38*, 7911–7920. [CrossRef]

63. Chae, J.; Joo, S.-K. Demand Response Resource Allocation Method Using Mean-Variance Portfolio Theory for Load Aggregators in the Korean Demand Response Market. *Energies* **2017**, *10*, 879. [CrossRef]

64. Yemshanov, D.; Koch, F.H.; Lu, B.; Lyons, D.B.; Prestemon, J.P.; Scarr, T.; Koehler, K. There is no silver bullet: The value of diversification in planning invasive species surveillance. *Ecol. Econ.* **2014**, *104*, 61–72. [CrossRef]

65. Stirling, A. Diversity and ignorance in electricity supply investment: Addressing the solution rather than the problem. *Energy Policy* **1994**, *22*, 195–216. [CrossRef]

66. Stirling, A. *On the Economics and Analysis of Diversity*; Science and Technology Policy Research (SPRU) Electronic Working Paper Series; Paper No. 28; University of Sussex: Brighton, UK, 1998.

67. Kruyt, B.; van Vuuren, D.P.; de Vries, H.J.M.; Groenenberg, H. Indicators for energy security. *Energy Policy* **2009**, *37*, 2166–2181. [CrossRef]

68. Charwand, M.; Gitizadeh, M.; Siano, P. A new active portfolio risk management for an electricity retailer based on a drawdown risk preference. *Energy* **2017**, *118*, 387–398. [CrossRef]

69. Karandikar, R.G.; Khaparde, S.A.; Kulkarni, S.V. Quantifying price risk of electricity retailer based on CAPM and RAROC methodology. *Int. J. Electr. Power Energy Syst.* **2007**, *29*, 803–809. [CrossRef]

70. Rohlfs, W.; Madlener, R. Assessment of clean-coal strategies: The questionable merits of carbon capture-readiness. *Energy* **2013**, *52*, 27–36. [CrossRef]

71. Cheng, C.; Wang, Z.; Liu, M.; Chen, Q.; Gbatu, A.P.; Ren, X. Defer option valuation and optimal investment timing of solar photovoltaic projects under different electricity market systems and support schemes. *Energy* **2017**, *127*, 594–610. [CrossRef]

72. Copiello, S.; Gabrielli, L.; Bonifaci, P. Evaluation of energy retrofit in buildings under conditions of uncertainty: The prominence of the discount rate. *Energy* **2017**, *137*, 104–117. [CrossRef]

73. Shimbar, A.; Ebrahimi, S.B. The application of DNPV to unlock foreign direct investment in waste-to-energy in developing countries. *Energy* **2017**, *132*, 186–193. [CrossRef]

74. Zhang, G.; Du, Z. Co-movements among the stock prices of new energy, high-technology and fossil fuel companies in China. *Energy* **2017**, *135*, 249–256. [CrossRef]

75. Broadstock, D.C.; Filis, G. Oil price shocks and stock market returns: New evidence from the United States and China. *J. Int. Financ. Mark. Inst. Money* **2014**, *33*, 417–433. [CrossRef]

76. Schaeffer, R.; Borba, B.S.M.C.; Rathmann, R.; Szklo, A.; Castelo Branco, D.A. Dow Jones sustainability index transmission to oil stock market returns: A GARCH approach. *Energy* **2012**, *45*, 933–943. [CrossRef]

77. De-Llano, F.; Iglesias, S.; Calvo, A.; Soares, I. The technological and environmental efficiency of the EU-27 power mix: An evaluation based on MPT. *Energy* **2014**, *69*, 67–81. [CrossRef]

 *energies*

Article

The Conundrum of Carbon Trading Projects towards Sustainable Development: A Review from the Palm Oil Industry in Malaysia

Tengku Adeline Adura Tengku Hamzah [1], Zainorfarah Zainuddin [1,*], Mariney Mohd Yusoff [1], Saripah Osman [1], Alias Abdullah [2], Khairos Md Saini [1] and Arno Sisun [1]

[1] Department of Geography, Faculty of Arts and Social Science, University of Malaya, Kuala Lumpur 50603, Malaysia; adelineadura@um.edu.my (T.A.A.T.H.); mariney@um.edu.my (M.M.Y.); saripahosman@um.edu.my (S.O.); khairos_ms@siswa.um.edu.my (K.M.S.); arno_sisun@yahoo.com (A.S.)

[2] Department of East Asian Studies, Faculty of Arts and Social Science, University of Malaya, Kuala Lumpur 50603, Malaysia; dralias@um.edu.my

* Correspondence: zainorfarah.zainuddin@gmail.com; Tel.: +60-12-3923948

Received: 22 July 2019; Accepted: 6 September 2019; Published: 13 September 2019

Abstract: Palm oil's utilization as a renewable energy (RE) source has led the government to intervene by introducing emission reduction projects. Carbon trading projects are part of the strategic direction adopted within the climate mitigation plan and sustainability drive in the palm oil industry. The perquisites and opportunities encountered within emissions trading are expected to aid palm oil producers economically, environmentally, and socially. This study addresses and analyses how the carbon trading projects' targets in Malaysia can be achieved, the problematic, and pressing issues around their implementation and whether these projects are sustainable and create a positive impact. This paper is based on literature reviews and semi-structured interviews with expert palm oil producers in Malaysia. The findings have revealed that carbon trading implementation in Malaysia has delivered new insights towards the international climate policy approach on the feasibility and impact of long-term sustainability goals. However, the impact of the implementation needs support from the government for further development. In conclusion, the major contribution of this study is that the carbon trading implementation in Malaysia complies with the objectives and principles of sustainable development and creates a significant influx in investment for Malaysia's economic growth.

Keywords: carbon trading; palm oil producer; renewable energy; economy; sustainable development; clean development mechanism; Sustainable Development Goals (SDGs); Malaysia

---

1. Introduction

In recent years, the literature on emissions trading regulation has expanded rapidly. This is due to the awareness relating to environmental change and a sustainable future gaining more attention worldwide. The global debates on climate change and mitigation strategies have led policymakers to address the government concerns about promulgating and implementing regulations to overcome the challenges of mitigating carbon emissions [1] and be more resilient in overcoming climate impacts. Malaysia has ratified the United Nations Framework Convention on Climate Change (UNFCCC) under the Kyoto Protocol since 2002. Malaysia's commitment to the carbon market is also closely related to, and affected by, overall climate change policy developments. Since the Paris Agreement in 2015, the UNFCCC has annually improved the climate policy standard, which also directs the international carbon market.

The carbon trading implementation review has changed drastically post-Kyoto and has developed into the new sustainable development mechanism (SDM) and sustainable development goals (SDGs).

The United Nations (UN) SDGs have set out a vision for future sustainability principles and Malaysia is one of the countries that has committed to engage in their implementation. Malaysia's national economic development policies, which were put into practice more than forty years ago, reflected several UN SDGs even before the introduction of the initiative in 2016 [2]. For example, the ambitious goals and targets that Malaysia set have been indicated through sustainable and systematic efforts by putting into practice the mechanism for the SDGs. Despite the desired sustainable goals, the palm oil industry in Malaysia plays a significant role in influencing growth towards trade and industry globalization business environments [3]. Moreover, Malaysia has acted on various strategies to invest in sustainability projects, as well as complying with sustainable practices in the palm oil business.

The clean development mechanism (CDM) is one example of a carbon trading project. The mechanism involves value-added investments that can generate carbon credits from environmental projects that have a monetary value. It is reviewed under the UN SDGs initiatives and its main aims are, first, to achieve sustainable development and productive environmental practices for present and future generations; second, to conserve the country's uniqueness and diverse cultural and natural heritage with effective participation by all sectors of society; third, to achieve sustainable living and strategies in sustainable production [4,5]. The CDM serves not only as an instrument to fight climate change issues but also as an important tool to achieve rapid economic growth for sustainable development and a strategy that can provide a constant source of business opportunities for corporations. CDM is designed to support foreign direct investment and clean technology transfers, leading to a reduction in greenhouse gas (GHG) emissions.

The palm oil industry in Malaysia has taken the necessary steps to respond to technological advancements as well as having carried out research. The growing upstream sector has had a rising impact on the whole value chain [6]. Based on various studies, as indicated, the carbon market plays a distinct role in bringing together various concerns across the globe and growth between developed countries and underdeveloped nations, reflecting opportunities that might result in a win-win situation and the achievement of the carbon market instrument [7,8]. However, few studies have been conducted to understand the challenges that have led to a decrease in carbon trading implementation in Malaysia. Furthermore, carbon trading ought to result in sustainable development and lead to benefits, such as investment, technology transfers, poverty relief, development of rural areas, and the provision of new opportunities for developing expertise and knowledge [9,10], more so in developing nations, but there has been a lack of further research indicating the impact of carbon trading on sustainable development.

A number of lessons can be learned from the CDM experience in the new policy framework's development. As in Figure 1, Malaysia is one of the biggest palm oil producers in the world, thus, the review of the findings suggests that the government needs to view the role played by the markets in a new way. The policy formulated must not simply be a repetition of past practices and must also be given time to make proper contributions to achieve the projects' goals. In other words, there must be a shift from offsetting and there should be greater concentration to be issued for financial verification, as well as an integration of human rights, public consultations, and transparency, as the mechanism's core principles. This mitigation action on emission reduction will provide useful recommendations that can facilitate the achievement of an efficient and robust SDM in the future.

In analyzing the possible future of palm oil sustainability and carbon trading, the significant benefits for the development in rural areas and biodiversity need to be examined in order to identify the projects' challenges and their sustainable development impact. The challenges and factors that have previously confused palm oil producers have continued to concern the palm oil industry in recent years. Despite this challenge, the engagement level and follow-up actions have continued to rise. However, with the European Union's (EU) threat of banning palm oil use in biofuels being perceived as trade discrimination, the carbon trading scheme appears to have an uncertain future.

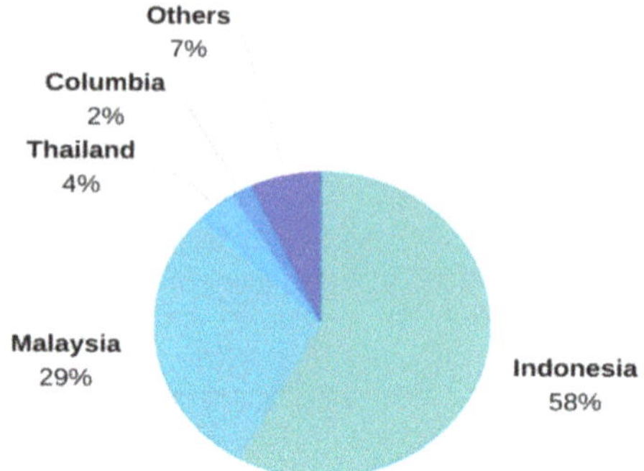

Figure 1. Palm oil production by country, 2017 [11].

The paper aims is to discover the challenges that have led to the conundrum of carbon trading projects and whether the palm oil companies that have engaged with carbon trading have achieved sustainable development from the projects. This research is intended to facilitate an incisive understanding of future development in palm oil companies in Malaysia to ensure a better standard of practices relating to climate change mitigation strategies and would contribute to the SDGs and national development. In addition, based on the qualitative research in this study and the analysis of the palm oil companies that are engaged with carbon trading in Malaysia, this research unpicks the carbon trading policy towards developing better strategies to allow palm oil to meet the sustainability goals.

2. Literature Review

2.1. Carbon Trading in Malaysia

The biomass renewable energy (RE) project's implementation in the carbon market is a cooperative mechanism that depicts high innovation under the bioenergy policy agenda. The implementation is planned with the intention of providing aid to developing countries to achieve sustainable development and to comply with climate change mitigation strategies [1]. The implementation is also aimed at assisting the developed world to comply with their commitment to reducing GHG emissions [12] and provide new opportunities in investments, technology transfer, building skills, and knowledge to create a sustainable future in Malaysia.

The CDM project activities have become one of the mitigation and emission reduction strategies in Malaysia and other countries in Southeast Asian regions, such as Singapore, Thailand, Indonesia, and the Philippines [12]. These countries are also engaged in carbon emission policies relating to the SDG standard and are focused not only on the energy sector but also on the agricultural and forestry sectors. In the Association of Southeast Asian Nations (ASEAN) countries, the CDM has been very successful in engaging with RE project application and GHG reduction [13]. As the carbon trading mechanism can be employed in separate projects, its effectiveness in encouraging RE use in developing nations, particularly in Asia, has a great potential impact on the energy industry.

In the palm oil industry, the Malaysia Palm Oil Board (MPOB) has continually prioritized research and development (R&D) to enhance sustainability in relation to palm oil. Moreover, the issues related to sustainability and productivity are core elements of consideration for the MPOB, consisting of various strategies to maximize the connection between economic development and environmental sustainability. Hence, the palm oil industry's future growth is based on a sustainable framework and the adoption of innovative technologies.

Malaysia's government has put considerable efforts into RE's development in the palm oil industry and into finding an alternative approach through various support and promotion programs to address the new carbon pricing instruments. The Nationally Appropriate Mitigation Actions (NAMAs) is one of the alternatives for supporting the CDM activities' continuity. Malaysia's palm oil industry has endeavored to implement projects that are relevant to RE production, such as RE supply, biofuel, biomass, biodiesel, and other bio-related production [14]. RE is known as the most adaptable type of low carbon and sustainable power source because it can contribute towards long-term emission reduction within the energy usage of electricity, transportation, and energy intensity [15]. The ratification in the UNFCCC discusses and implements the strategy for reducing carbon emissions [16] and one of the strategies noted was to have businesses innovate and invest in carbon trading projects that can be applied in the country's palm oil industry.

2.2. CDM in the Palm Oil Industry

As [17] mentions, the palm oil industry in Malaysia has great potential to be engaged in carbon trading projects. The amount of biomass and palm oil production has been noted to increase from time to time [18] and, therefore, it assumes a strategic role in enhancing RE consumption and delivering a sustainable future. The RE production from palm oil mill effluent (POME), biomass, composting, and bioenergy residues can prove beneficial for different sectors in moving forward to a new market mechanism. This goal is one of the directions for carbon trading project implementation.

As of 2015, there were registered CDM projects with a total investment of approximately USD1,530 million in certified emission reductions (CER) transactions and holdings in Malaysia's pending account, with 2,789,528 CER [5]. The biomass production accounted for 80.1% of Malaysia's CDM pipeline or 76.9% of all registered projects [14]. The data on the CDM project undertakings in Malaysia shows a massive biomass production volume from milling and plantation activity [19]. The data for 2015 indicates that project activities involved processing oil palm excesses, biofuel, biomass, methane capture, and co-composting using either solid or liquid waste collectively. Table 1 provides a breakdown of the projects' undertakings and their distribution in the palm oil industry according to project type.

Table 1. 2015. Project undertakings (clean development mechanism (CDM)) and their distribution in palm oil producers by project type [5,14].

Type	Subtype	Total	Registered Projects	Registered Projects with Issuance	At Validation
Biomass energy	Palm oil solid biomass	33	31	11	2
	Agricultural residues	5	5	2	-
	Wood waste	4	4	-	-
	Gasification	1	1	-	-
CH4 Avoidance	POME	60	54	18	6
	Composting	36	27	6	9
	Total	139	122	37	17

* POME = Palm oil mill effluent.

The CDM projects' status in Malaysia has been given due consideration due to the sharp decline in the carbon market following uncertainty in the future global carbon market [14]. According to the data by the Ministry of Natural Resources and Environment (MNRE) [14], 35.8% of Malaysia's RE-based CDM project pipelines are from biomass energy. Most of the palm oil mills generated biomass by capturing methane gas from POME treatment and used it as a fuel source for generating electrical and thermal energy for either on or off-site consumption. The Palm Oil National Key Economic Areas (NKEAs) plan, for instance, targets achieving 100% compliance by all the mills that have implemented biogas recovery projects by 2020. Moreover, this initiative has put in place a policy designed to re-operationalization all CDM methane avoidance projects in the palm oil sector [14]. In addition, the relevant public policies, including the National Renewable Energy Policy Action Plan (2010) and the Economic Transformation Programme (ETP), are focused on enhancing the utilization of local renewable energy sources as a way of contributing towards the security of the nationwide supply of electricity and to sustainable socio-economic development.

According to the 2015 Malaysian government report [20], the nationally determined mitigation contributions (NDCs) demonstrate the government's commitments to minimize the intensity of GHG emissions by 45% by 2030 relative to the intensity of emission of gross domestic product (GDP) in 2005. The amount of GHG emissions is made up of 35% on a non-restricted basis and an additional 10% as the condition upon proof of payment of climate funding, transfer of technology, and capacity building from developed countries [20]. As indicated in Table 2, the biennial report [21] made available by the MNRE gives a summary of the reduction of emissions as of 2013 and possible emission reduction in 2020. The comparison in emission reductions between 2013 and 2020 shows a huge reduction in emissions with the palm oil industry mitigation action. The statistics suggest that specific mitigation actions have been put in place to ensure the nation meets it's target in the NDCs report to the UNFCCC.

Table 2. List of achieved emissions reductions in 2013 and projected for 2020 [21].

Sector	Mitigation Action	Emission Reduction Achieved in 2013 ($ktCO_2eq$)	%	Potential Emission Reduction in 2020($ktCO_2eq$)	%
Energy	RE implementation through the FiT mechanism	252.78	1.36	5458.09	16.96
	RE electricity generation by FiT regulated public and private licensees and other mechanisms	948.77	5.1	2179.29	6.77
	Blended petroleum diesel in biodiesel (palm-based)	719.74	3.87	1802.49	5.60
	Green technology application	94.81	0.5	1426.35	4.43
	Green building rating scheme implementation	60.40	0.3	858.40	2.67
	Efficient electricity consumption in all Federal Government ministry buildings (baseline established in 2013)	-	-	98.21	0.30
	Development and usage of energy-efficient vehicles (EEVs) in emissions reduction	40.96	0.22	199.74	0.62
	Compressed natural gas (CNG) use in motor vehicles	154.62	0.83	217.57	0.68
	Rail-based public transport	214.93	1.16	977.51	3.03
LULUCF	Sustainable forest management	13,797.37	74.26	13,800.00	42.88
Waste	Wastepaper recycling	1993.47	10.72	2159.45	6.71
	Biogas captured from palm oil mill effluent (POME) treatment	300.95	1.62	3001.89	9.32
	Total	18,578.80		32,178.99	

* FiT= Feed-in Tariff.

With the potential development in carbon trading projects, the CDM projects' CER in Malaysia have contributed a total of 9,844,435 CER issued as of April 2015 [14]. However, the CER oversupply has reduced its price and it is unable to meet the high transaction costs incurred in verification. As discussed in the Conference of Parties 21 (COP 21) in Paris, carbon trading projects require longer-term outcomes to ensure optimal growth in sustainability; hence, developing countries can grow economically and achieve their respective emission targets with a low economic cost. Other studies have focused on how well carbon trading projects function; however, the numerous prospects for initiating CDM project activities in the palm oil industry will demonstrate significant achievements in the future [13]. Furthermore, the subsidies provided for the CDM's financial allocation have received criticism from many organizations who have contended that the allocation has been subject to unethical and unsuitable implementation in reducing carbon emissions. Based on various sources and the detailed exploration by the Standards and Industrial Research Institute of Malaysia (SIRIM) and the Malaysian Palm Oil Council (MPOC) [22], the slow-moving carbon trading and engagement in Malaysia within the palm oil industry is because of the lack of knowledge and uncertain development within the carbon market and trading opportunities [23,24], which have led to various questions and dilemmas in the carbon market system involving the palm oil industry in Malaysia. However, as presented in Table 3 below, the total CER issued are higher within the palm oil industry. The development towards the SDGs aims to enhance renewable energy development in Malaysia and would bring material and positive outcomes to the palm oil industry [14].

Table 3. Potential and certified emission reductions in Malaysia's CDM pipeline in the palm oil industry [14].

Type	Sub-type	Registered Projects (ktCO$_2$eq/year)			Projects at Validation (ktCO$_2$eq/year)	Total (ktCO$_2$eq/year)
		Annual ER	Expected Accumulated ER up to 2012	Total CER Issued	Annual ER	Annual ER
Biomass Energy	Palm oil solid biomass	2547.431	8642.982	5604.858	349.551	2896.982
	Agricultural residues	615.834	3456.930	114.472	-	615.834
	Wood waste	110.777	163.896	0.000	-	110.777
	Gasification	26.983	35.968	0.000	-	26.983
	POME	2249.808	4511.034	1411.135	205.425	2455.233
CH4 Avoidance	Composting	770.107	2756.419	203.207	808.732	1578.839

3. Research Methodology

As described above, this study focuses on capturing the outlook of critical factors in carbon trading implementation in the palm oil industry. The issues of carbon trading challenges are crucial yet problematic. As such, a qualitative study was conducted to understand these issues, adopting one-to-one interviews and discussions. This methodology developed as a qualitative approach and a connecting strategy for scientific research and local knowledge [25].

3.1. Data Collection

As identified by [26], the quantity of data that can be utilized that is gathered from the interviewees and the number of participants required for the interviews has a converse relationship. The criterion for selection for the interviewees was purposive suitable sampling. After the selection of the various experts in carbon trading in the palm oil industry, information seeking the prospective interviewees' permission to engage in semi-structured interviews was channeled through email or telephone messages. As highlighted above, due to the limitations in some carbon trading experts, a small number of them were interested in taking part in a follow-up study via an interview. The researcher deliberately included carbon trading firms in the palm oil industry to accomplish the research objectives in terms of size and segment.

In particular, the interview questions explored the respondents' perception of the challenges and dilemmas that could affect the carbon trading implementation growth. The data collected was based on literature reviews, current status, growth, challenges and trends, and government and private sector reports were considered as the secondary data collection process determinants. The primary interview data collection was proposed in this study to provide an explanation of how the respondents replied to the carbon trading project issues in the palm oil context. Semi-structured interviews were conducted and they offered a more in-depth understanding of the focus subject. These interviews were directed by the researcher and were digitally recorded after the interviewees gave their verbal agreement. Furthermore, the researcher took notes required at the time of interview.

A total of seven company representatives were interviewed for data collection for this study. The sampling frame was drawn from the UNFCCC database, totalling 49 palm oil companies in Malaysia. The interviews were conducted with professionals among carbon trading experts and institutional theory was adopted to structure the interview questions. This theory informs the combination of deductive and inductive logic. The analyzed data was compared and developed further based on the data collected from the interviews. The researcher applied thematic analysis for the purpose of identifying a variety of restricted themes under every category, which could adequately reflect the data that has been collected. Thematic analysis is the qualitative data encoding procedure under a variety of themes. The two main ways through which themes can be identified in the thematic analysis include the inductive and deductive methods [27]. The inductive method is strongly linked with the data itself during the process of identification [28], while the deductive method entails the use of themes that are guided by the theoretical or analytical interest in the area, as well as a more explicitly analyst-driven approach.

In the current research, an investigation of four main categories was carried out, namely the legal dimension, the financial conditions, the green resources and capabilities and the positive impacts on SDM. Under each group, a few themes were generated inductively based on the raw information obtained from the subjects. A compilation of the identified themes is available in the research findings section.

3.2. Data Analysis

In this paper, the focus on the palm oil companies is according to the detailed content development and data analysis information presented in the companies' sustainability reports or other equivalent corporate publications. The following sections introduce and discuss the analysis of the respective evaluation forms from the supplementary literature reviews, desktop reviews, and interviews. Table 4 shows the profile of the firms that took part in the data collection, consisting of the distribution in years of establishment, location, and the respondents/interviewees' positions as representatives of Malaysian palm oil companies.

Table 4. Respondent's profiles.

Company(Code)	Years of Establishment	Geographical Location	Position
A	30	Perak	Manager
B	25	Penang	Senior Manager
C	42	Pahang	Manager
D	24	Johor	General Manager
E	36	Selangor	General Manager
F	46	Perak	Manager
G	32	Kedah	Researcher

A transcription of the interviews was then carried out and the coded data was further analyzed through the thematic analysis. As illustrated by [29], qualitative analysis is often conducted through reviewing the subjects' answers to every question and then carrying out a selection of the most relevant parts to the question asked. The participants had expressed various conditions that are perceived to be significant in facilitating their views on carbon trading based on the experiences they have. The factors were grouped into four broad themes and were subsequently mapped to the four forms of assessment for pressure from institutions that were anticipated in this research. Table 5 below presents the four themes and their relevant descriptions.

This study has explored the positive impacts of CDM implementation within palm oil companies in Malaysia. Its findings have supported the significant factors on the legal dimensions, financial conditions, green resources, and capabilities, and highlighted the more important criteria that could encourage the palm oil industry to implement CDM, which has positive impacts on sustainable development. Based on the follow-up interviews' exploratory investigation, some palm oil companies have confirmed that the majority of Malaysian palm oil producers have considered implementing CDM as their future option in relation to sustainability and can engage in long-term investment. However, there are some limitations in relation to its achievement as the future carbon market development has an effect on global policy and, at some points, education, skills, technology, and cost factors interfere with the decisions.

Table 5. Interview findings on crucial factors and positive sustainable development impacts.

Categories	Sub-Categories	Companies						
		A	B	C	D	E	F	G
Legal dimension	Not compatible with national regulations	*			*		*	
	Difficult to engage with international policy				*	*		*
	Strict additionally in emissions projects		*	*		*	*	
Financial conditions	Low confidence level in financial returns		*			*	*	*
	The carbon market's uncertainty	*	*	*	*	*	*	
	Lack of financial investment			*				*
	Difficult CER sales and higher consultancy costs		*		*		*	
Green resources and capabilities	Lack of green resources in clean technology infrastructures					*		
	A requirement for new technical skills			*				
Positive impacts the on Sustainable Development impact in palm oil Industry	Environmental, social, and economic	*	*		*	*	*	*

* Supported findings.

4. Discussion

This section aims to discuss the results obtained in this work and to describe the main contribution of this research. As can be observed in Table 5, many of the factors show the challenges and crucial factors in carbon trading projects within the palm oil industry in Malaysia. Elaborating on these crucial factors is important in addressing the developed research is necessary to verify the challenges of carbon trading implementation.

4.1. Legal Dimension

The existing CDM framework is seen to be lacking in terms of guidelines for improving sustainability performance [30]. Most of the CDM projects involved in development or positioning in RE technologies have been criticized for their negative impact on the environment and social fields. Supported by the results in Table 5 from the palm oil companies, the Malaysian palm oil industry particularly offers a promising platform for CDM projects. However, proper monitoring needs to be performed in order to ensure the projects' positive impact. Furthermore, a set of sustainability criteria should be developed to promote the achievement of sustainable development and the criteria should reflect the local context in which the projects are carried out. In other words, the effort must be holistic in order to reach an equilibrium state for sustainability.

The legal dimension factor's results showed that three out of seven respondents indicated "not compatible with national regulations." In Malaysia, there are no regulations to dictate that the palm oil industry should implement carbon trading. Carbon trading implementation, such as CDM, was made as a result of climate negotiations and, therefore, governments with different views on carbon trading's commodification influence the outcome [31].

The analysis of the collected data revealed that the majority of the interviewees on carbon trading implementation in Malaysia considered the "legal dimension" as the crucial factor in implementing carbon trading. As argued by a representative from Company A, the regulatory constraints hinder the implementation of CDM and the development of the carbon market, which indicates the perception of the crucial legal dimension factors in carbon trading implementation (Company A, 2018).

However, Companies B, C, E, and G considered the legal dimension as one of the incentives to meet the sustainable palm oil standard and development in palm oil plantation areas. As stated by [32], the firms respond to the institutional pressure that emphasizes the importance of the regulatory, normative, and cognitive factors that have affected the firms' decisions to adopt some environmental practices. Beyond the institutional pressures' perspective, perhaps more symbolic actions have been a long-standing priority for palm oil companies to provide emission-related investments and to participate in voluntary initiatives that allow them to demonstrate their commitment to the government as core corporate strategy elements on climate change [33].

The results relating to the "difficulty to engage with international policy" show that Companies D, E, and G have affected the management decisions to continue in CDM projects. The representative from Company E (2018) argued that the project developers and consultants are essentially dependent on each other and, if they don't comply with each other and there is a problem, then the consultant needs to be changed. This was explained further by the representative from Company G (2017), "There are a lot of problems, and they don't want to share the loan equally. You are the industrialized country, I am also the industrialized country, the ratio that we want to split is not agreeable."

There are many arguments that have advocated carbon trading implementation related to the environmental market and policies governance. While governance is important in policy formation, the critical factor on strict additionality in implementing carbon emission projects has contributed to confusion in continuing with the projects. From the findings, "strict additionality in emissions projects" shows that four of the seven palm oil companies agreed with the problem. As stated by representatives from Company B (2018), "Review the entire procedures of the data monitoring, collections and compilation and documentations ... once wrong, another problem will start ... typical accounting procedures."

Moreover, when climate change became a big issue, there were two different sorts of arguments that economists used to explain efficient methods for reducing emissions [34]. Although the world has recognized Malaysia as being ready to replace non-renewable energy with renewable fuels, the implementation of various policies and programs by the Malaysian government has increased the awareness of the important role of RE in sustainable energy systems [35]. The existing policy, however, appears to not be indicative of such progress.

The ongoing revision from the Malaysian perspective has contributed to the development of a mechanism that complies more sustainably with the palm oil industry. More importantly, the enhancement of carbon market policies has shown a positive impact. This analysis provides links between the legal dimension perspectives with the current Malaysian policies. The challenge of addressing carbon trading stresses the palm oil industry's potential to contribute to several SDGs, especially for sustainable palm oil, RE, waste management, and water treatment of palm oil waste, which appear to be beneficial for promoting sustainable development.

4.2. Financial Conditions

Carbon trading projects in Malaysia have shown a downturn since 2015 [14]. Project developers have shown less interest and confidence in this project's implementation. According to the MNRE Consultancy Report [14], the issues hampering the continuity is that there has been a lack of demand but an increase in the CER supply. The findings identified that the "low confidence level and uncertainty of the carbon market" has contributed to the growth of financial flows and resources. There is also a low commitment to further investment to continue the current depressed market situation. Nevertheless, the challenge must be met in order to persist with Malaysia's pledge to voluntarily reduce its emissions by up to 45% by 2030.

For the palm oil companies, one of the barriers to investing in carbon trading projects is the high costs of CER consultation and verification. The weakening of the local currency (consultancy and verification fees are generally quoted in €), paired with the drop in CER value, has made CDM projects a costly endeavor (Company B, 2018).

Furthermore, as highlighted by representatives of Companies E and F, the challenge in the current carbon trading implementation, specifically in the Malaysian palm oil industry, concerns the registration and certification costs. For palm oil producers, GHG reduction projects are implemented to reduce the overall carbon footprint of the palm oil production itself. Hence, the trading of any carbon credits defeats this purpose (from a carbon accounting position) unless it has exceeded sustainability targets (Company F, 2018).

As mentioned, Malaysia has benefited from the investment in GHG emissions reduction projects through the ratification in the UNFCCC. As the representative from Company B (2018) pointed out, " ... investment in renewable energy is costly and political support for the adoption of renewable energy has been inconsistent and unfair." When the regulators removed the subsidies, the carbon market's inconsistencies caused the carbon prices to fluctuate [14] and affected the demand for carbon. Therefore, CDM projects can only earn the return on investment (ROI) by continuance to invest in the longer term despite focusing on the uncertainties on the mechanism and other new mechanisms.

The representative from Company C (2017) argued that there was a "lack of financial investment primarily influenced by demand and supply ... and low-prices due to low-demand." As mentioned by [31], there was a drop in CDM prices because, after Kyoto, the markets were demanding fewer credits and the financial crisis caused problems. The issues of concern are that "difficulty of CER sales are higher in consultancy" generate an additional cost for the project developers. As pointed out by [36], the cost is related to the environmental projects' formalization and validation, the monitoring process and the implementation verification process. However, costing and CER sales are dependent on the size of the project and requires a huge investment of the project development. Companies A, C, E, and G disagree on the "difficulty of CER sales and higher consultancy costs" because the credits received met their expectations and satisfied the financial additionality criteria of CDM.

The financial benefits from the sale of carbon credits have been viewed as being potentially positive and economically profitable. Therefore, environmental practices create advantages for palm oil companies to benefit from emissions reduction and reduced energy intensity or by switching to relatively cheaper fuels as part of the operational costs.

4.3. Green Resources and Capabilities

In Malaysia, many options under green technology projects could create significant co-benefits in this country. The strategies are able to address local and regional environmental problems and advance social goals [37]. For developing countries that might otherwise give priority to their immediate economic and environmental needs, the prospect of significant additional benefits should provide a strong inducement to participate in carbon trading projects.

In this study, the terms of green resources and capabilities have been identified as being able to minimize energy and resource dependence and to make economic trends more sustainable with resource-efficient manufacturing [38]. In order to meet the growing demand for palm oil in the future, the palm oil producers must co-exist with new skills development and green resources to move towards sustainability.

In theory, irrespective of the allocation of the rights to trade in the carbon market, Malaysia benefits from a transaction that is designed to contribute towards sustainable development. Similar to other foreign investments, the RE projects in Malaysia could promote technology transfer and the adoption of similar strategies and policies for emissions reduction planning. In contrast, green resources and capabilities do not support the "lack of green resources in green technology" and "need new technical skills."

From the findings, the representative from Company C (2017) stated that "The carbon trading project presents an opportunity for channelling of resources towards projects that are most likely to be accessed for further national sustainable development." In addition, new skills development is needed to move forward towards sustainability as one of the emerging knowledge and new skills (Company A, 2018). The analysis suggests that, if the management systems and skills development are not well developed, the external pressures on sustainability issues will be relatively weak and unresponsive [33]. Thus, responding to sustainability-related projects, such as carbon trading, limits a company's ability to move forward.

Consistent with the findings in this study, carbon trading promotes green resources and capabilities, which were significant in the previous study by [39]. One of the CDM objectives was to promote green technology and increase local skills and equipment content [39]. Carbon trading can also provide employment, new knowledge skills and business opportunities in less developed countries (Company B, 2018).

There are high expectations amongst palm oil producers in Malaysia due to the potential of carbon trading to deliver sustainable benefits and its ability to attract foreign investments, technology transfer, and possible contributions to poverty alleviation [40]. With initiatives in energy sectors and bioenergy production, the palm oil carbon emission projects' activities have already produced the benefits of sound progress that can contribute to sustainable development. Carbon trading projects can lower the cost of compliance with the emissions reduction initiative projects for developed countries.

4.4. Sustainability in the Palm Oil Industry

The primary purpose of carbon trading is to reduce overall global carbon emissions. The concept behind carbon trading is simple: if total global emissions are reduced, it does not matter from which country or organization the reduction comes. Thus, this concept will benefit developed nations and organizations by enabling them to purchase CER to meet their NDC targets (Company B, 2018).

Based on the findings for the sustainable development's positive impact, it is shown that six of the seven palm oil companies have stated a positive sustainable impact. The representative from Company C argued (2017), "In certain countries, there has been criticism that carbon trading does not promote sustainable development because the activities act as an enabler for developed countries to maintain and increase their emission levels as long as they have the means to pay in order to meet their set targets."

Furthermore, despite the issues associated with carbon trading in the palm oil industry, sustainable palm oil's evolution [41] in Malaysia should be perceived as being more effect in contributing ideas to more sustainable reductions in carbon emissions. However, in relation to Malaysia's economic growth, it can be concluded that the growth of new palm oil estates in the tropical rainforest will soon no longer be achievable [9].

In the mitigation's development process, it is vital to consider the quality of the environment. In practice, there exists a significant strain between financial and environmental targets in fulfilling the carbon emissions projects. An ethical concern faced in this industry is that the production of palm oil affects the environment, wildlife, and native communities. Despite the arguments about sustainability issues concerning carbon trading, Malaysia has established policies related to sustainability that ensure economic development and ecological biodiversity. By considering an alternative approach and supplying the economic and technological undertakings, numerous issues could be removed [5]. Additionally, the resources that are already on-going are more advanced and are using technologies with good practices; however, the resources need to be scaled up and used in other appropriate ways [42].

The SDGs are influential, long-term prospects that can provide the direction and unlimited framework for businesses driven towards sustainability to release their powers of transforming and developing solutions for the world. In short, the SDGs are driving businesses to become an even stronger force for good. A recent report by [43] has compared the potential of implementing carbon trading projects and has asserted that most companies are defined by carbon reduction benefits, as well as the social benefits that are associated with the palm oil sector. As discussed, sustainable development benefits the reduction of air and water pollution through the minimized usage of fossil fuel, as well as extending to improving water availability, reducing soil erosion, and the protection of biodiversity. From the perspective of social benefits, many projects would result in employment opportunities in the different regions or income groups that are targeted and would result in local energy self-reliance promotion. The goals of carbon reduction and sustainable development can, therefore, be pursued on a long-term basis.

Carbon trading projects also have an impact on funding channels in achieving economic, social, environmental, and sustainable development targets, including clean water and atmosphere, that sit alongside social development, such as growth in rural areas, creation of job opportunities, alleviation of poverty, and often reduce the reliance on the use of imported fossil fuels [44]. Alongside green investment opportunities in developing nations, the voluntary carbon trading projects in the palm oil industry also improve climate and future developments, as well as environmental issues within local areas. The prospect of such benefits ought to provide a strong inducement for developing countries that are anxious about immediate economic and social development to engage in palm oil carbon emissions and sustainable future projects.

5. Limitations and Future Research

This paper acknowledges that there are limitations to this study as it focuses only on palm oil producers in Malaysia and the findings illustrate the challenges and sustainable development impact within the palm oil industry. However, these findings are important in determining how palm oil producers will enhance the strategies and future policies in this field. In another context, the alternatives in carbon trading project implementation should be implemented locally and lead to the improvement in future carbon emissions projects.

The second limitation pertains to palm oil producers in the country which have conventionally struggled to come to terms with the ban on biofuels produced using palm oil as well as the deforestation ban phenomena which began in the early part of 2018. For this reason, the number of participants was limited, and the majority of producers rejected the invitation of the interview. Moreover, earlier on in 2019, the trade war that broke out with palm oil producers curtailed the producers' ability for participation. Furthermore, the involvement in this study was both subjective and voluntary; for this reason, it is difficult to reach the subject matter expert in carbon trading and CDM.

It is necessary to broaden the scope of this study to encompass other divergent viewpoints, including economic dimensions, stakeholders' perspectives, international relations, social responsibility, as well as competitiveness in future research. Finding novelty in other perspectives and industries could also help improve the level of understanding and be in alignment with aspirations relating to Malaysia's SDGs. This approach could also be used for a quantitative method or mixed methods in future order to gather detail for future studies. Furthermore, studies in the future must undertake research on the energy industry, which could turn out to be one of the most promising sectors to enforce carbon trading as well as contributing toward sustainable development.

Malaysia has been proactive in accelerating the adoption of the 2030 Agenda and SDGs nationally. The country began preparing for the SDGs since 2014 in order to incorporate the SDGs into the national planning framework with a view to reduce the emissions target by 2030. The CDM or carbon trading was the tool that was used previously. Currently, there is no encouragement to continue with the projects and it is just business-as-usual. Through the implementation of carbon trading, Malaysia can maintain its flexibility in adjusting and realigning strategies based on the achievements, challenges, and lessons learned during the previous phase, while also assessing emerging trends and circumstances that may affect the desired development outcomes [43]. It also provides an opportunity for feedback and greater participation from various stakeholders as they gradually align themselves with the 2030 Agenda as well as the SDGs.

However, the research work for future studies would yield better results with an evaluation and examination of the study model in other industries so that comparisons can be made with the existing research work. Hence, value would be added concerning the insight to find the appropriate enforcement regulations in carbon trading, as well as the completion level and market demand for the local market with carbon trading implementation.

6. Conclusions

In recent years, sustainability has been promoted by the addition of carbon trading as a vital strategy for an agricultural business. Palm oil is one of the biggest traded commodities and has provided many benefits to health, society, the economy, and the environment. Many policies have emerged in Malaysia with regard to achieving the SDGs' agenda. The palm oil industry's potential in carbon trading project utilisation in Malaysia could be improved with government encouragement in the carbon market post-2020.

This paper has explored carbon trading implementation challenges in Malaysia that will assist in leading the opinions on problems and solutions for a better transition. In the struggle over palm oil's sustainability, Malaysia has shown a strong desire to promote itself as a significant palm oil hub in the South-East Asia region, with extra payback from carbon trading implementation. Thus, the government has to play a more active role. The government can improve conditions for local development by providing funding opportunities and policies that could increase carbon trading's competitiveness. Supportive policies need to be developed to enhance the implementation and investment from the private sector in the agricultural sector, particularly in the palm oil industry. The new trend of business opportunities has accelerated the palm oil sector's development towards a sustainable future and improved process efficiencies. Despite the challenges to be solved, the on-going R&D into palm oil carbon trading projects are expected to promote a more advanced generation of palm oil bioenergy and biotechnology, where high value-added products and bio-based chemicals are produced, and technology enhancements are provided for the palm oil producers. In this study, carbon trading projects' positive sustainability impacts have contributed to new intervention in the palm oil industry towards more sustainable production, with improvements in procedures and policies, alongside the SDGs agenda. Through technology transfer and engagement in carbon trading projects, Malaysia might meet its target in reducing the regional temperature to 1.5 degree Celsius, in accordance with the Paris Agreement pledge that the targeted carbon emissions would be reduced.

This paper offers an original contribution to the research field in carbon trading within the palm oil industry. It aims to improve the understanding of palm oil producers and policymakers for future developments in carbon trading and its impact through contributing to a sustainable future with opportunities to drive sustainable development and the climate change mitigation strategies in order to achieve SDGs. By understanding the challenges in carbon trading implementation, this study hopes to further develop and enhance new approaches and to encourage continuous improvement in environmental projects that soon will become the new period of future development. Moreover, the challenges relating to the difficulty in implementing the CDM will help policymakers, practitioners, and other industries to understand and manage future climate mitigation strategies.

The concept of sustainability has pinned a promising outlook for the palm oil industry to continue the development and growth of palm oil use. Despite the challenges with the legal dimension, the financial conditions and the green resources and capabilities, the palm oil industry still shows a significant inflow of investments and has been following Malaysia's extensive economic growth. Furthermore, Malaysia will be able to utilize its substantial green potential by developing a robust supply network for proper palm oil supply and demand connections in the carbon market. The potential of the mitigation action objectives will enhance the sustainability performance in the palm oil industry in Malaysia, in line with the SDGs.

Author Contributions: T.A.A.T.H. and Z.Z. designed and developed the paper; Data Analysis and Formatting was conducted by S.O. and K.M.S.; writing—original draft preparation was conducted by M.M.Y.; writing—review and editing was conducted by Z.Z.; visualization was conducted by A.A. and A.S.; supervision was conducted by T.A.A.T.H. and A.A.

Funding: This paper is supported by the University of Malaya—(UMRG) SUSCI Grant no. RP024A-17SUS and IPPP Grant PG247-2015B.

Conflicts of Interest: The authors declare no conflict of interest.

References

1. Owusu, P.A.; Samuel, A. A Review of Renewable Energy Sources, Sustainability Issues and Climate Change Mitigation. *Cogent Eng.* **2016**, *3*, 1167990. [CrossRef]

2. The Star. Malaysia Makes Good Progress in Adopting UN Global Goals. Available online: https://www.thestar.com.my/business/business-news/2018/05/09/msia-makes-good-progress-in-adopting-un-global-goals/#kMSPVtxLs83hYi5T.99 (accessed on 30 December 2018).

3. Bessant, J. Youth Participation: A New Mode of Government. *Policy Stud.* **2003**, *24*, 87–100. [CrossRef]

4. Lehmann, H.; Rajan, S.C. *Sustainable Lifestyles: Pathways and Choices for India and Germany*; Deutsche Gesellschaft: Berlin, Germany, 2015.

5. UNDP. Malaysia. UNDP Report. 2015. Available online: http://www.my.undp.org/content/malaysia/en/home/sustainable-development-goals/goal-13-climate-action.html (accessed on 19 February 2018).

6. MPOB. Foreword from the Director General Malaysian Palm Oil Board. 2015. Available online: http://www.mpob.gov.my/en/about-us/foreword-from-director-general (accessed on 30 January 2018).

7. Matsuo, N. CDM in the Kyoto Negotiations. *Mitig. Adapt. Strateg. Glob. Chang.* **2003**, *8*, 191–200. [CrossRef]

8. Michaelowa, A.; Dutschke, M. Integration of Climate and Development Policies through the Clean Development Mechanism. In *Europe and the South in the 21st Century. Challenges for Renewed Cooperation*; Eadi, G., Ed.; Karthala: Paris, France, 2002.

9. United Nations Office for Sustainable Development (UNOSD). *Achieving Climate Change Action and the Sustainable Development Goals*; UNOSD: Lilongwe, Malawi, 2018.

10. Kaupp, A.; Liptow, H.; Michaelowa, A. CDM is Not about Subsidies—It is About Additionally. *Energise* **2002**, *1*, 8–9.

11. Santhia, V. Essential Palm Oil Statistics 2017. Palm Oil Analytics. Available online: www.palmoilanalytics.com/flles/epos-final-59.pdf (accessed on 30 December 2017).

12. UNEP. *Sustainable Development Priorities for Southeast Asia*; UNEP RRC.AP.: Pathumthani, Thailand, 2004.

13. Amran, A.; Zainuddin, Z.; Zailani, S.H.M. Carbon Trading in Malaysia: Review of Policies and Practices. *Sustain. Dev.* **2012**, *21*, 183–192. [CrossRef]

14. MNRE. *Review of Clean Development Mechanism Activities as Potential Nationally Appropriate Mitigation Actions*; UNDP Consultancy for the Ministry of Natural Resources and Environment Malaysia: Putrajaya, Malaysia, 2015.

15. Bakar, D.A.; Anandarajah, G. Sustainability of Bioenergy in Malaysia Concerning Palm Oil Biomass Adopting Principles Governing Bioenergy Policy in the UK. *Energy Sustain. V Spec. Contrib.* **2015**, *206*, 57–67.

16. MITI. MITI Report 2015. Miti.gov.my. Available online: http://www.miti.gov.my/miti/resources/MITI_Report_2015-5.pdf (accessed on 8 May 2019).

17. MPOB. Oil Palm Planted Area by Category. 2012. Available online: http://bepi.mpob.gov.my (accessed on 30 January 2018).

18. Shafie, S.M.; Mahlia, T.M.I.; Masjuki, H.H.; Andriyana, A. Current Energy Usage and Sustainable Energy in Malaysia: A Review. *Renew. Sustain. Energy Rev.* **2011**, *15*, 4370–4377. [CrossRef]

19. Shuit, S.H.; Tan, K.T.; Lee, K.T.; Kamaruddin, A.H. Oil Palm Biomass as a Sustainable Energy Source: A Malaysian Case Study. *Energy* **2009**, *34*, 1225–1235. [CrossRef]

20. UNFCCC. Intended Nationally Determined Contribution of the Government of Malaysia. 2015. Available online: https://www4.unfccc.int/sites/submissions/INDC/Published%20Documents/Malaysia/1/INDC%20Malaysia%20Final%2027%20November%202015%20Revised%20Final%20UNFCCC.pdf (accessed on 16 January 2019).

21. MNRE. *Biennial Update Report to the UNFCCC*; UNFCCC: Putrajaya, Malaysia, 2015.

22. MPOC. Promoting a Sustainable Way Forward for Oil Palm Cultivation. 2017. Available online: http://www.mpoc.org.my/ (accessed on 18 January 2018).

23. Eng, A.G.; Zailani, S.; Abd Wahid, N. A Study on the Impact of Environmental Management System (EMS) Certification towards Firms' Performance in Malaysia. *Manag. Environ. Qual. Int. J.* **2006**, *17*, 73–93. [CrossRef]

24. Daddi, T.; Frey, M.; De Giacomo, M.R.; Testa, F.; Iraldo, F. Macro-Economic and Development Indexes and ISO14001 Certificates: A Cross National Analysis. *J. Clean. Prod.* **2015**, *108*, 1239–1248. [CrossRef]

25. Cornwall, A.; Jewkes, R. What is Participatory Research? *Soc. Sci. Med.* **1995**, *41*, 1667–1676. [CrossRef]

26. Morse, J.M. Strategies for sampling. In *Qualitative Nursing Research: A Contemporary Dialogue*; Morse, J.M., Ed.; Sage: Newbury Park, CA, USA, 1991; pp. 127–145.

27. Frith, H.; Gleeson, K. Clothing and Embodiment: Men Managing Body Image and Appearance. *Psychol. Men Masc.* **2004**, *5*, 40–48. [CrossRef]

28. Patton, M. *Qualitative Evaluation and Research Methods*; Designing Qualitative Studies; Sage: Beverly Hills, CA, USA, 1990; pp. 169–186.
29. Talja, S. Analyzing Qualitative Interview Data: The Discourse Analytic Method. *Libr. Inf. Sci. Res.* **1999**, *21*, 459–477. [CrossRef]
30. Kua, H.W. Improving the Clean Development Mechanism with Sustainability-Rating and Rewarding System. *Prog. Ind. Ecol. Int. J.* **2010**, *7*, 35–51. [CrossRef]
31. Bondevik, S. Carbon forestry and trading: A case study of Green Resources in Uganda. Master's Thesis, BI Norwegian Business School, Oslo, Norway, 2013.
32. Delmas, M.; Toffel, M. Institutional Pressures and Environmental Management Practices. In Proceedings of the 11th International Conference of the Greening of Industry Network, San Francisco, CA, USA, 12–15 October 2003.
33. Sullivan, R.; Gouldson, A. The Governance of Corporate Responses to Climate Change: An International Comparison. *Bus. Strategy Environ.* **2017**, *26*, 413–425. [CrossRef]
34. Paterson, M. Selling carbon: From international climate regime to global carbon market. In *The Oxford Handbook of Climate Change and Society*; Oxford University Press: Oxford, UK, 2011.
35. Mohamed, A.R.; Lee, K.T. Energy for Sustainable Development in Malaysia: Energy Policy and Alternative Energy. *Energy Policy* **2006**, *34*, 2388–2397. [CrossRef]
36. Michaelowa, A.; Hoch, S. *FIT for Renewables? Design Options for the Green Climate Fund to Support Renewable Energy Feed-In Tariffs in Developing Countries*; World Future Council: Hamburg, Germany, 2013.
37. Lim, X.L.; Lam, W.H. Review on Clean Development Mechanism (CDM) Implementation in Malaysia. *Renew. Sustain. Energy Rev.* **2014**, *29*, 276–285. [CrossRef]
38. European Commission. *Communication from the Commission to the European Parliament, the Council, the European Economic and Social Committee and the Committee of the Regions*; Roadmap to a Resource Efficient Europe: Brussels, Belgium, 2011.
39. Winkler, H.; Spalding-Fecher, R.; Mwakasonda, S.; Davidson, O. Sustainable development policies and measures: Starting from development to tackle climate change. In *Building on the Kyoto Protocol: Options for Protecting the Climate*; World Resource Institute: Washington, DC, USA, 2002.
40. Olsen, K.H. The Clean Development Mechanism's Contribution to Sustainable Development: A Review of the Literature. *Clim. Chang.* **2007**, *84*, 59–73. [CrossRef]
41. Ivancic, H.; Koh, L.P. Evolution of Sustainable Palm Oil Policy in Southeast Asia. *Cogent Environ. Sci.* **2016**, *2*, 1195032. [CrossRef]
42. EPU. Malaysia Sustainable Development Goals Voluntary National Review 2017. High-level Political Forum. Economic Planning Unit. Available online: https://sustainabledevelopment.un.org/content/documents/15881Malaysia.pdf (accessed on 3 September 2019).
43. UNFCCC. Katowice Climate Change Conference—December 2018. Available online: https://unfccc.int/process-and-meetings/conferences/katowice-climate-change-conference-december-2018/katowice-climate-change-conference-december-2018 (accessed on 15 July 2019).
44. UNFCCC. Available online: http://cdm.unfccc.int/index.html (accessed on 4 December 2018).

 *energies*

Article

The Importance of Heat Emission Caused by Global Energy Production in Terms of Climate Impact

Anna Manowska * and Andrzej Nowrot

Department of Electrical Engineering and Automation in Industry, Faculty of Mining, Safety Engineering and Industrial Automation, Silesian University of Technology, Akademicka 2, 44-100 Gliwice, Poland
* Correspondence: anna.manowska@polsl.pl

Received: 20 June 2019; Accepted: 6 August 2019; Published: 9 August 2019

Abstract: The global warming phenomenon is commonly associated with the emission of greenhouse gases. However, there may be other factors related to industry and global energy production which cause climate change—for example, heat emission caused by the production of any useful form of energy. This paper discussed the importance of heat emission—the final result of various forms of energy produced by our civilization. Does the emission also influence the climate warming process, i.e., the well-known greenhouse effect? To answer this question, the global heat production was compared to total solar energy, which reaches the Earth. The paper also analyzed the current global energy market. It shows how much energy is produced and consumed, as well as the directions for further development of the energy market. These analyses made it possible to verify the assumed hypothesis.

Keywords: global heat production; energy market; energy conversion; electricity

---

1. Introduction

Our daily life on Earth requires the production of large amounts of energy. The energy is produced mainly in the forms of electrical energy and mechanical energy as a result of liquid fuels and gases combustion in engines in various types of vehicles. Unfortunately, the various types of useful energy also cause the production of a huge amount of heat. Moreover, these useful energies are eventually converted to heat energy in machines, vehicles, and devices. An important question arises here—whether this release of a large amount of thermal energy is comparable to solar energy and can it significantly affect the global temperature and climate change. In this paper, the thesis was verified whether the heat emission, which is the final result of energy generated by our civilization, has an effect comparable to that of greenhouse warming. The greenhouse effect is well known and is comprehensively presented in many papers—among others [1–4]. However, the effect and greenhouse gas emission have not been discussed in the presented work so far.

Nuclear and fossil fuel power plants are thermal power stations and have a maximum efficiency of around 40%. It means that during the production of electrical energy, around 60% of nuclear or chemical energy in fossil fuel is converted to heat energy and is emitted directly into the atmosphere and water (seas, lakes, rivers). Thus, most of the energy contained in any fuel is lost, but there is currently no technically more efficient way to generate electricity. Unfortunately, that also applies to the combustion of biofuels. Wind and water power plants have much higher energy efficiency, and the production of electricity is accompanied by a low heat emission. Much more information about the energy conversion efficiency in thermal power stations is presented in the papers [5–7].

However, regardless of the way how electricity is produced, almost all electricity is ultimately turned into heat. It may seem surprising, but the following analysis proves this thesis. Electrical energy is distributed to different electricity consumers—factories, buildings, hospitals, railway electrification systems, etc. For example, any computer consumes electrical power and uses it to perform calculations, which is accompanied by heat production in semiconductor components (microprocessors, transistors, diodes, resistors). In practice, considering the field of physics, all the electric power in any computer is finally converted to heat. In a fridge, an electric motor drives a heat pump, which pumps out heat from inside the fridge to the outside. According to the principle of energy conservation, the heat energy released into the atmosphere (outside) is equal to the sum of heat energy collected from the fridge interior and the consumed electricity. A very similar situation occurs in any device. While an electrical train moves, electric motors cause acceleration driving at a constant velocity and partial energy recovery during braking. When the velocity of the vehicle is constant, all the consumed electrical energy compensates the work of friction forces in mechanical components (in bearings, etc.) and of air resistance forces. The friction in the components causes their heating, and air resistance causes the heating of the train's surface and air. The same situation occurs in cars. In a modern petrol engine, the chemical energy in the fuel is converted to heat (about 60%) and useful mechanical energy (about 40%). All mechanical energy is used to compensate friction forces and air resistance, so always a heat is produced. Thousands of various devices and machines can be considered in which the energy would finally be converted into heat—similar to what it is shown in Figure 1.

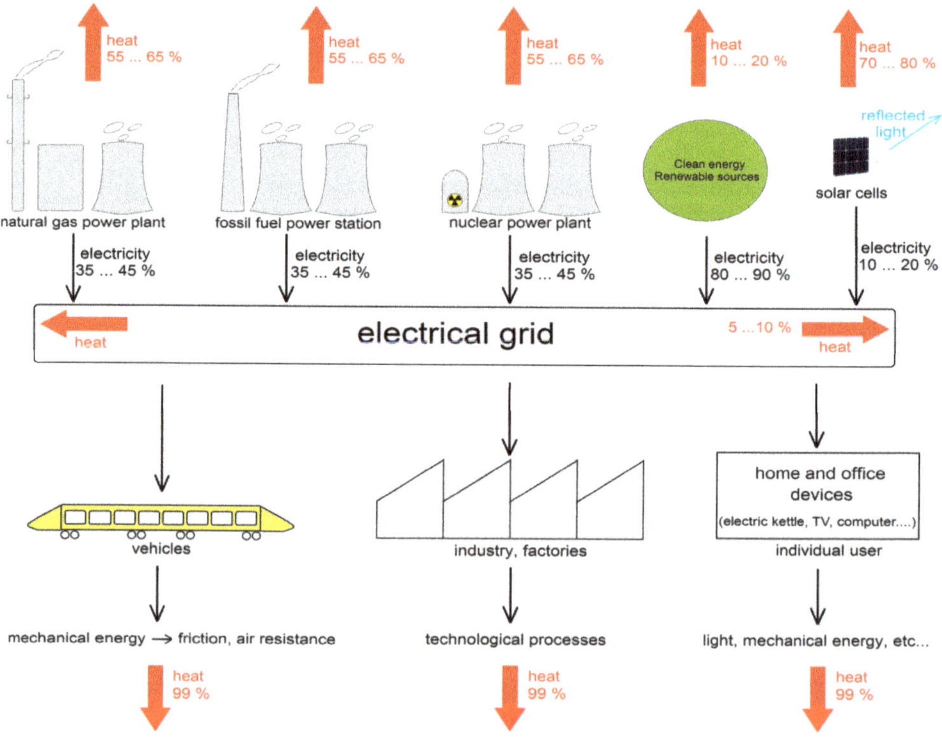

Figure 1. Electrical energy changes into other forms of energy and makes useful work, but almost 100% is finally converted to heat energy (own study).

Consequently, sooner or later, almost 100% of electrical energy or chemical energy in fuels is finally converted to heat energy. To determine the global heat production, it is necessary to calculate the global production of electricity and petrol fuels.

2. Energy Market

The right energy structure and the increase in energy efficiency are the key issues on the way to the transformation in the field of energy, which is to ensure a safe future for all [8–11]. The current way of generating energy is associated with high greenhouse gas emissions because it is based mainly on fossil fuels and the inefficient use of traditional bioenergy, especially in developing countries [12–16]. Besides, the world's demand for energy is growing, which is largely due to the economic growth that drives global energy consumption. Rapid growth occurs in developing countries, such as China and India, while in countries belonging to the OECD (The Organisation for Economic Co-operation and Development), for example, Germany or Italy, energy consumption is beginning to stabilize but is still at a high level. These tendencies are in line with the assumptions of the Singer report [17], and this can also be confirmed by analyzing current trends, both in consumption and in energy production, as shown in Figure 2.

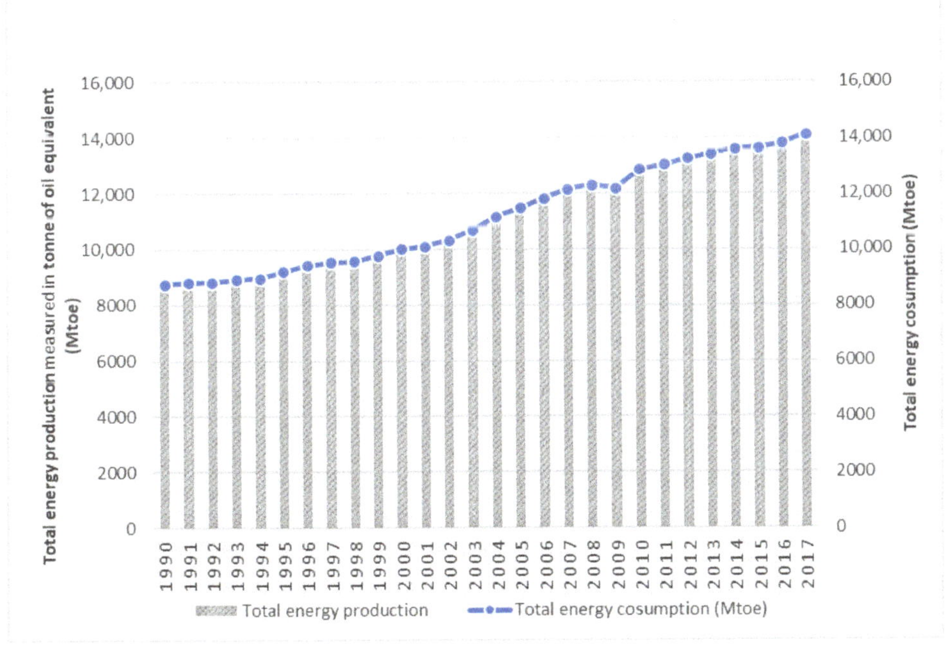

Figure 2. Total energy consumption and production, source: own elaboration based on data [18].

Let us first take a look at how global energy production has changed from a long-term perspective, both in terms of quantity and sources (Figure 3).

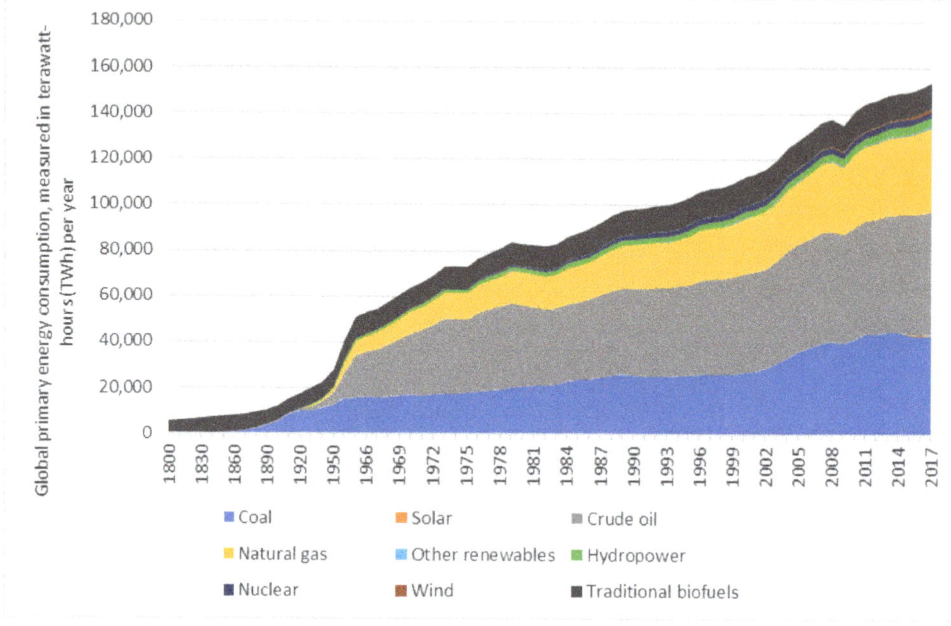

Figure 3. Global primary energy consumption 1800–2017 [19].

In the 1800s, almost all of the world's energy was produced by burning wood and other organic matter [19]. Oil consumption started in the 1870s [19]. Twenty years later, there were natural gas and hydroelectricity in the energy mix. Since 1900, coal consumption has increased significantly, and, now, it constitutes almost half of the world's energy [19]. In the 20th century, the energy mix changed significantly, coal took over traditional biofuels, and oil constituted up to around 20% of the energy market, and since 1960, the world has had nuclear electricity [18]. Today, on the energy market, we have renewable energy sources—RES, not appearing until the 1980–1990s [19].

Besides, apart from analyzing the total consumption of primary energy, we could also compare the contribution of different energy sources. As shown in Figure 4, changes in the energy mix are slow, so, mainly, fossil fuels dominate [19]. Except for the emergence of nuclear electricity, the energy mix was fairly stable for at least fifty years [18].

Changes in the structure of the energy mix began after 1997 [20,21] when the first EU regulations on energy from renewable sources appeared [22–25]. They were stated in the European Commission's White Paper Energy for the Future—Renewable Energy Sources RES (December 1997). In 2001, the European Parliament and the Council adopted Directive 2001/77/EC on the promotion of electricity production from RES on the internal market, which determined the share of RES in total electricity consumption by 2010 (replaced by Directive 28/2009/IN). A breakthrough in the development of RES took place in March 2006, after the publication of the Green Book [26].

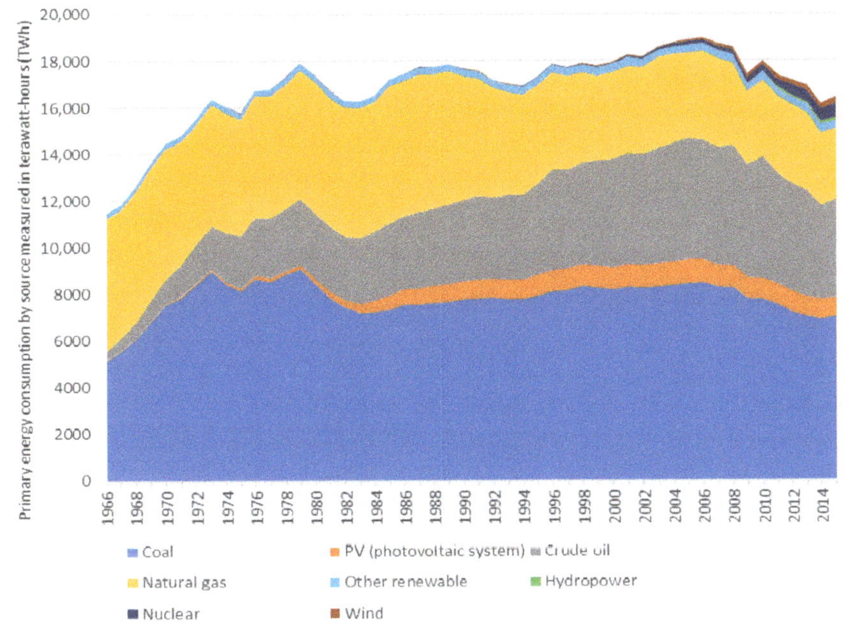

Figure 4. Primary energy consumption by source, European Union [19].

Figure 5 shows primary energy consumption from 1965–2017 in continental regions [19].

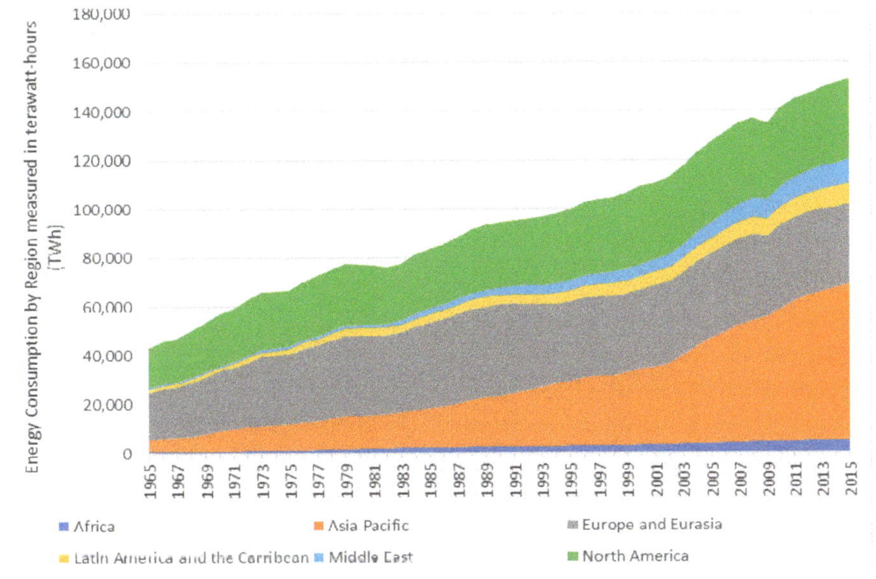

Figure 5. Global energy consumption by region [19].

In 1965, most of the total energy was consumed in North America, Europe, and Eurasia. In total, they accounted for over 80% of global energy consumption. Consumption in other parts of the world has been increasing, most dramatically in the Asia Pacific. In 2015, Asia Pacific was by far the largest regional consumer. The total consumption (about 43%) was the same as in North America, Europe, and Eurasia combined [19]. The Middle East used 7%, Latin America 5%, and Africa 3% [18,19].

In 2017, the production of energy from coal increased due to China. Higher energy prices in the world had caused a drop in oil and gas production in the United States. Energy production in the European Union decreased due to moderate growth in energy consumption, lower primary energy production (nuclear and hydro), depletion of oil and gas resources, and climate policy, which ultimately means giving up coal. Large oil and gas exporting countries, such as Russia, Iran after international sanctions, Canada or Nigeria, as well as fast-developing countries (India, Indonesia, Turkey, and Brazil), were the main energy donors. In 2017, total consumption was at 567 EJ (EJ = exajoule = 10^{18} joules), and it is forecast that by 2020, there would be a 10% increase [19].

Prices can have a big impact on the choice of energy sources. While comparing costs between sources, it is important to have relative prices [19]. In the literature, the levelized cost of electricity (LCOE) is used [27]. The concept of LCOE is defined in the following way:

- it measures the cost of living divided by energy production,
- it calculates the current value of the total cost of construction and operation of the plant during the assumed period of use,
- it allows comparison of different technologies (e.g., wind, solar, natural gas) with uneven life span, project size, different capital costs, risk, return, and capacity [27].

Figure 6 shows the levelized cost of electricity (LCOE). The cost range is represented on the column chart for each technology. The white line in each of them represents the global cost of each technology. The average range of fossil fuel costs is shown as a gray column. In this chart, it is shown that, in 2016, most renewable technologies were in a competitive range to fossil fuels. The key exception was the thermal solar radiation, which remained about twice as expensive, but began to decrease [19]. Hydropower is the oldest renewable source. In this technology, the price is low. Analyzing how the average cost of technology changed in the years 2010–2016, we can see that the cost of solar photovoltaic system dropped significantly [19]. This reduction of cost in photovoltaic cells has been deep over the past few decades. The price of solar photovoltaic modules has decreased more than 100 times since 1976 [19].

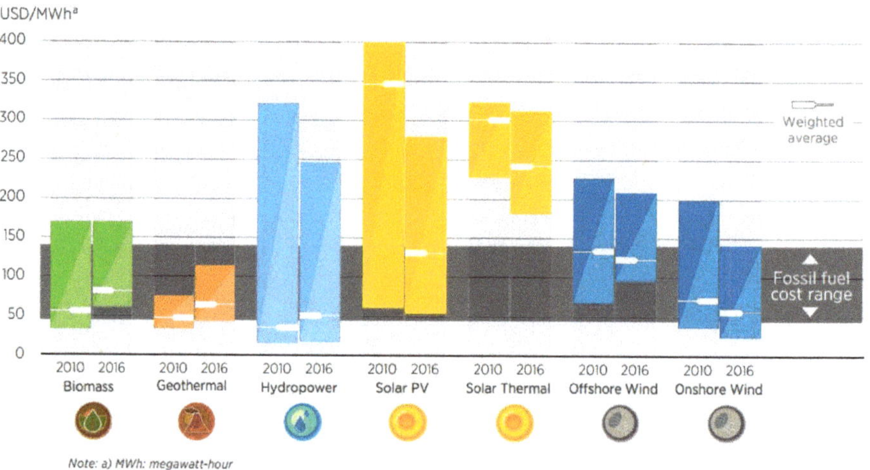

Figure 6. Levelized cost of electricity (LCOE) in 2010 and 2016 [19].

Measuring the number of people with electricity access is a very important social and economic indicator. We think that the growth of population has caused an increase in energy consumption. Figure 7 shows that the percentage of people with access to electricity has increased at a global level [19].

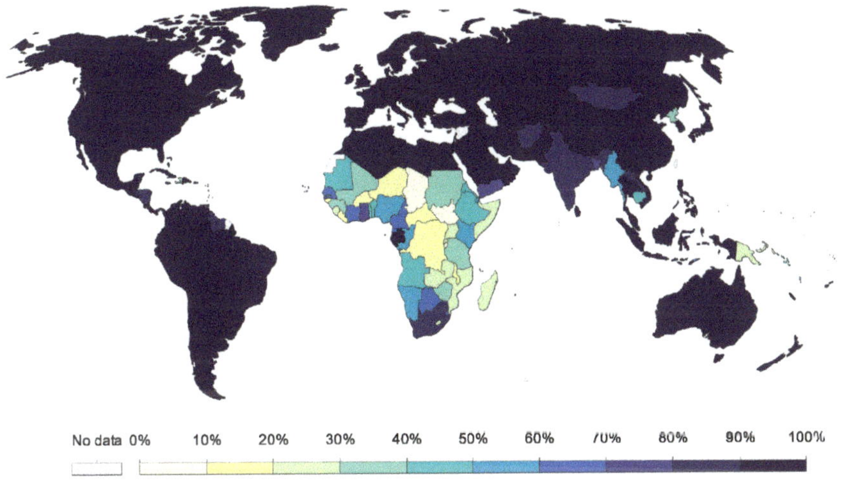

Figure 7. Share of the population with access to electricity [19].

In 1990, about 73% of the population had access to electricity, and this level increased to 85% in 2014 [19]. High-income countries typically have access between 95–100% [19]. The growing global share was, therefore, driven by increased access in low- and middle-income economies. In many countries, this trend has developed, for example, access in India has increased from 45% to 80%, and in Indonesia, it has increased to 97% from 60% in 1990 [19]. While the trend is upward for most countries, there are some countries where this access is very low, for example, only 8.8% of Chad's population has electricity access. Therefore, we can assume that this factor will increase, and thus also energy consumption, in the future.

3. Materials and Methods

The structures of the energy market and total energy production are important to determine the global heat emission. While producing useful energy, as it is shown in the introduction, heat energy is emitted (as an adverse side effect)—Equation (1). Moreover, sooner or later, almost 100% of that useful energy—electrical energy or chemical energy in fuels—is finally converted to heat energy. Because average efficiency of a thermal electric plant is about 40% [5–7], most of the primary energy (chemical energy in coal or natural gas, nuclear energy in nuclear fuel, etc.) is converted directly to heat. Engines in cars, where chemical energy in petrol is converted to mechanical energy, have similar efficiency. The results in Figure 2 show that the values of global energy production and energy consumption are very similar—they are approximately the same. Because the global annual useful energy production is known (Figures 2 and 3), taking the efficiency about 40%, total annual energy produced on the Earth could be written by Equations (1) and (2). Unfortunately, as it is shown in Figure 3, the amount of energy produced by renewable sources is still very low on a global scale. Therefore, it is reasonable to adopt the average efficiency of the energy production process at 40%, as in thermal power plants, because this dominates. The total power produced by our civilization on the Earth is determined by Equation (3).

$$E_{total} = E_{heat} + E_{useful} \tag{1}$$

$$E_{useful} \approx 0.4 \times E_{total} \tag{2}$$

$$P_{total} = \frac{E_{total}}{3600 \times 24 \times 365} \tag{3}$$

where E_{useful}—annual useful energy production (electricity, motion, etc.)—it is given in Figures 2 and 3, E_{total}—total annual energy, which is converted to heat (as an adverse side effect) and useful energy, E_{heat}—annual heat energy production, P_{total}—total power produced on the Earth (in Watts). All energies are in joules per year.

Based on the global energy production of useful energy (E_{useful}), it is possible to estimate the total energy (heat energy and useful energy), which is also finally converted to heat in consumers. At this point, the question arises whether the value of the calculated E_{total} (in global terms) is high, low, significant, or insignificant. To answer this question, the total power produced on the Earth would be compared to the total solar irradiation reaching the Earth's surface. It is presented by factor n in Equation (4).

$$n = \frac{P_{total}}{P_{GND}} \tag{4}$$

where n—the ratio of total power produced on the Earth to the power of solar irradiation hitting the Earth's surface, P_{GND}—solar irradiation power hitting the Earth's surface (the ground).

The heat produced on the Earth combines with the energy supplied from the Sun. Consequently, these two energies together cause the energetical effect as if the Earth was a little closer to the Sun. To illustrate this effect, we imagined the Earth before the industrial revolution (before the 19th century) in orbit with a radius of r_0—Figure 8. Nowadays, additional heat emission causes an effect of the seeming shortening of the Earth's orbit to a radius of r_1.

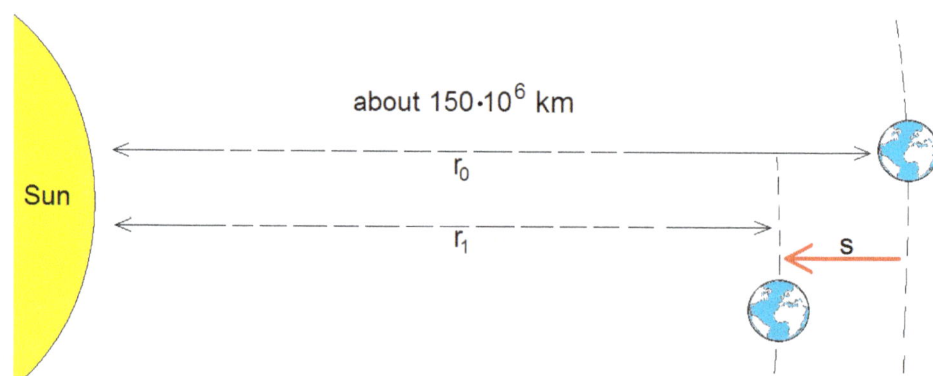

Figure 8. Global heat emission causes the seeming energetical effect as if the Earth was a little closer to the Sun.

The seeming shortening of the Earth's orbit is given by Equation (5).

$$s = r_0 - r_1 \tag{5}$$

Sun radiance, it means the radiant power per unit area emitted over all wavelengths and all directions (watts per square meter) of the Sun's surface area, is given by the Stefan-Boltzmann law [8]—Equation (6). The total power emitted by the Sun is given in Equation (7).

$$R^* = \sigma \times T^4 \tag{6}$$

$$P_{Sun} = R^* \times A_{Sun} \tag{7}$$

where:

R^*—Sun radiance—power emitted by one square meter of the Sun's surface area (W/m^2)
P_{Sun}—total power emitted by the Sun (Watts)
A_{Sun}—Sun's surface area (m^2)
T—the temperature of the Sun's surface
σ—Stefan-Boltzmann constant $5.67 \cdot 10^{-8}$ (W m^{-2}K^{-4})

The sunrays are emitted over all directions from the Sun. As we move away from the Sun (the distance from the Sun increases), the intensity of light decreases. For the distance "r" from the Sun, the solar radiation strikes the surface (sphere) whose area is equal to $4\pi r^2$. The surface power density (called also solar constant, flux density) for the distance "r" is represented by Equations (8)–(11).

$$P_d(r) = \frac{P_{Sun}}{4 \times \pi \times r^2} \tag{8}$$

$$P_d(r) = \frac{A_{Sun} \times \sigma \times T^4}{4 \times \pi \times r^2} \tag{9}$$

introducing the constant "k", we obtain:

$$k = \frac{A_{Sun} \times \sigma \times T^4}{4 \times \pi} \tag{10}$$

$$P_d(r) = k \times \frac{1}{r^2} \tag{11}$$

where P_d—power density, solar constant, power per unit of surface reached to place at a distance of "r" from the Sun in a unit of W/m^2, r—distance from the center of the Sun (in meters)

Finally, the surface power density for the point at a distance of "r" from the Sun is inversely proportional to the square of the distance. Because the mean distance of the Earth from the Sun is about 149.6×10^6 km [9], the surface power density above the atmosphere of the Earth is about 1380 W/m^2, theoretically based on Equation (9), and about 1366 W/m^2 in reality [10].

4. Results and Discussions

Global useful energy produced on the Earth (Figure 2) in 2000 was equal 10,016 Mtoe (419.35×10^{18} J), and in 2017 was equal 14,080 Mtoe (589.50×10^{18} J). Total produced energy (heat + useful energy) is given in Equation (12) (transforming Equation (2)).

$$E_{total} \approx \frac{E_{useful}}{0.4} \tag{12}$$

The Sun always illuminates the surface of the Earth seen as a wheel. Total power of the solar rays reaching the top of the Earth's atmosphere (Equation (13)) can be computed by multiplying the surface power density (Equation (9)) and illuminated Earth's surface—the wheel with a radius of 6378 km [9]. Very rare phenomena, such as the solar eclipse, are omitted.

$$P_{SE} = P_d \times \pi \times R_{Earth}^2 \approx 1.7 \times 10^{17},\ W \tag{13}$$

where R_{Earth}—radius of the Earth, P_{SE}—the power of the solar irradiation reaching the top of the Earth's atmosphere.

This result is confirmed by the Authors in work [11]. Around 8.1×10^{16} W of solar radiation passes through the atmosphere and arrives at the surface of the Earth, which accounts for 47% of the solar energy that reaches the Earth [11]—Equation (14).

$$P_{GND} \approx 0.47 \times P_{SE} \tag{14}$$

where P_{GND}—solar irradiation power hitting the Earth' surface (the ground).

The "n" ratio (total power produced on the Earth to the power of the solar irradiation ratio) is obtained in Table 1.

Table 1. Calculated values of P_{total} and "n" ratio. $P_{GND} = 8.1 \times 10^{16}$ W.

Year	E_{total}, J/Year	P_{total}, W	$n = \frac{P_{total}}{P_{GND}}$
2000	1048.37×10^{18}	3324×10^{10}	0.414×10^{-3} (0.0414%)
2017	1473.75×10^{18}	4673×10^{10}	0.577×10^{-3} (0.0577%)

Taking Equations (11) and (14), (15) could be written as:

$$P_{GND}(r) = 0.47 \times k \times \pi \times R_{Earth}^2 \times \frac{1}{r^2} \tag{15}$$

where $P_{GND}(r)$—solar irradiation power hitting the Earth's surface (the ground) as a function of the distance between the Earth and the Sun.

Next, we introduced the radius r_0 of the actual (real) orbit of the Earth and the hypothetical radius r_1 of the orbit, which is shorter and corresponds to a little higher solar power. It means the heat emission causes the effect of the seeming shortening of the Earth's orbit to a radius of r_1. Because the P_{GND} is inversely proportional to the square of the distance (Equation (16)), it is possible to determine the hypothetical radius r_1 (Equations (17)–(19)). Finally, the seeming shortening of the radius of the Earth's orbit is given by Equation (20).

$$P_{GND} \sim \frac{1}{r^2} \tag{16}$$

$$P_{GND1} = P_{GND0} + P_{total} = P_{GND0} + n \times P_{GND0} \tag{17}$$

$$r_1 = r_0 \times \sqrt{\frac{P_{GND0}}{P_{GND1}}} \tag{18}$$

$$r_1 = r_0 \times \sqrt{\frac{P_{GND0}}{P_{GND0} \times (1+n)}} = \frac{r_0}{\sqrt{n+1}} \tag{19}$$

$$s = r_0 - \frac{r_0}{\sqrt{n+1}} \tag{20}$$

where r_0—radius of the actual (real) orbit of the Earth, r_1—radius of a hypothetical orbit where the little higher solar power is the equivalence of the total energy produced on the Earth, s—seeming shortening of the radius of the Earth's orbit.

In conclusion, the heat emission as a consequence of energy production on the Earth gives the same effect as shortening the radius of the Earth's orbit by about 31,000 km in 2000 and about 43,300 km in 2017 with respect to the time directly before the industrial revolution (before the 19th century).

5. Conclusions

Total energy produced by our civilization on the Earth consists of the useful energy (electricity, mechanical energy, etc.) and heat, as an adverse side effect. Unfortunately, the various types of useful energy are finally converted to heat in machines, vehicles, and devices. The estimated calculations presented in this work prove that the total power produced by us was 0.0414% in 2000 and 0.0577% in 2017 of the global solar power reaching the Earth's surface—it means reaching the lowest layers of the atmosphere, almost the ground. A significant increase in the "n" factor between 2000 and 2017 results from the increase in global energy production.

A few questions arise here—are these values high or low? Are they important? A very good determiner of the importance of these values is the value of cyclic, annual changes of the Earth's orbit. The orbits of planets are not circles; they are ellipses. The eccentricity of planets is a very well-known natural phenomenon in the solar system. As a result of the elliptical trajectory of the Earth around the Sun, the solar light power density varies about 3.3% [11], and that is over 50 times more than the "*n*" factor in Table 1. After inserting data into Equation (20), the total heat produced on the Earth causes an effect like a seeming shortening of the radius of the Earth's orbit around the Sun by about 31,000 km in the year 2000, and about 43,300 km in the year 2017. Consequently, the seeming radius shortening ("*s*" in Equation (5), Figure 8) is about 2.4 (for 2000) and 3.4 (for 2017) times bigger in comparison to the Earth's diameter (12,756 km). These distances are significant compared to the Earth's dimensions, but, in the field of astronomy, they are very small. Taking into account that the average Earth-Sun distance is about 1496×10^5 km, the seeming shortening "*s*" is 3455 times smaller (in 2017). However, the key factor in assessing the importance of the heat emitted during the global energy production in terms of climate impact is the comparison of the parameter "*n*" to the natural changes of the solar light power density reaching the Earth. That power density varies about 3.3% cyclically during every year (the "*n*" is about 50 times smaller), and this is why the total heat energy produced by our civilization has a small impact on the global warming in the time horizon of several dozen years. We can say that there would be a one-year increase of supplied energy by about 3.3% of the average solar energy once every 50 years. It would probably have significance over several hundred years. Therefore, the use of renewable energy sources makes sense regardless of their energy efficiency because the emitted heat during the generation of electric power in such sources does not significantly affect the climate. For example, photovoltaic solar cells have a low energy efficiency—a dozen or so percent [28]. The fact that more than 80% of solar energy is converted into heat in photovoltaic panels should not be a limitation in their widespread use.

Author Contributions: Conceptualization, A.M. and A.N.; Methodology, A.M. and A.N.; Validation, A.M. and A.N.; Formal Analysis, A.M. and A.N.; Investigation, A.M. and A.N.; Resources, A.M. and A.N.; Data Curation, A.M. and A.N.; Writing—Original Draft Preparation, M.A. and A.N.; Writing—Review & Editing, M.A. and A.N.; Visualization, A.M. and A.N.; Supervision, A.M.; Project Administration, A.M.; Funding Acquisition, A.M.

Funding: This research received no external funding.

Conflicts of Interest: The authors declare no conflict of interest.

References

1. Qiao, H.; Zheng, F.; Jiang, H.; Dong, K. The greenhouse effect of the agriculture-economic growth-renewable energy nexus: Evidence from G20 countries. *Sci. Total Environ.* **2019**, *671*, 722–731. [CrossRef] [PubMed]

2. Jones, P.D. Greenhouse Effect and Climate Data. In *Reference Module in Earth Systems and Environmental Sciences*; Elsevier: Cambridge, MA, USA, 2013. [CrossRef]

3. Dunne, J.A.; Jackson, S.C.; Harte, J. Greenhouse Effect. In *Encyclopedia of Biodiversity*, 2nd ed.; Academic Press (Elsevier): Cambridge, MA, USA, 2013; pp. 18–32.

4. Kirk-Davidoff, D. The Greenhouse Effect, Aerosols, and Climate Change. In *Green Chemistry, An Inclusize Approach*; Elsevier: Amsterdam, The Netherlands, 2018; pp. 211–234.

5. Tillman, D.A. *Coal-Fired Electricity and Emissions Control Efficiency and Effectiveness*; Butterworth-Heinemann: Oxford, UK, 2018; pp. 3–299. ISBN 978-0-12-809245-3.

6. Bhatia, S.C. *Advanced Renewable Energy Systems*; Woodhead Publishing: Delhi, India, 2014; pp. 1–31, 490–508. ISBN 978-1-78242-269-3.

7. Fu, C.; Anantharaman, R.; Jordal, K.; Gundersen, T. Thermal efficiency of coal-fired power plants: From theoretical to practical assessments. *Energy Convers. Manag.* **2015**, *105*, 530–544. [CrossRef]

8. Kingston, R.H. Chapter 1—Blackbody Radiation, Image Plane Intensity and Units. In *Optical Sources, Detectors and Systems. Fundamentals and Applications*; Academic Press (Elsevier): Cambridge, MA, USA, 1995; pp. 1–32.

9. Basic Planetary Data. Available online: https://www.britannica.com/place/Earth/Basic-planetary-data (accessed on 1 June 2019).

10. Cohen, S.; Stanhill, G. Widespread Surface Solar Radiation Changes and Their Effects: Dimming and Brightening. In *Climate Change (Second Edition), Observed Impacts on Planet Earth*; Elservier: Amsterdam, The Netherlands, 2016; pp. 491–511.

11. Zheng, H. Chapter 2—Solar Energy Utilization and Its Collection Devices. In *Solar Energy Desalination Technology*; Elservier: Amsterdam, The Netherland, 2017; pp. 47–171.

12. Rybak, A.; Manowska, A. The forecast of coal sales taking the factors influencing the demand for hard coal into account. *Miner. Resour. Manag.* **2019**, *35*, 129–140.

13. Rybak, A.; Manowska, A. The role of hard coal in national energy security with regard to the energy policy in Poland until 2050. *Przegląd Górniczy* **2017**, *73*, 45–51.

14. Manowska, A.; Tobór-Osadnik, K.; Wyganowska, M. Economic and social aspects of restructuring Polish coal mining: Focusing on Poland and the EU. *Resour. Policy* **2017**, *52*, 192–200. [CrossRef]

15. Jonek-Kowalska, I. How do turbulent sectoral conditions sector influence the value of coal mining enterprises? Perspectives from the Central-Eastern Europe coal mining industry. *Resour. Policy* **2017**, *55*, 103–112. [CrossRef]

16. Brodny, J.; Tutak, M. Analysis of the diversity in emissions of selected gaseous and particulate pollutants in the European Union countries. *J. Environ. Manag.* **2019**, *231*, 582–595. [CrossRef] [PubMed]

17. Singer, S. *Demaskowanie Mitów: Obalanie Mitów o Energii Odnawialnej*; WWF Report: Gland, Switzerland, 2014; pp. 1–66.

18. Global Energy Statistical Yearbook. 2019. Available online: https://yearbook.enerdata.net/ (accessed on 1 June 2019).

19. Ritchie, H.; Roser, M. Energy Production & Changing Energy Sources. Available online: https://ourworldindata.org/energy-production-and-changing-energy-sources (accessed on 1 June 2019).

20. Kijewska, A.; Bluszcz, A. Research of varying levels of greenhouse gas emissions in European countries using the k-means method. *Atmos. Pollut. Res.* **2016**, *7*, 935–944. [CrossRef]

21. Bluszcz, A. Classification of the European Union member states according to the relative level of sustainable development. *Qual. Quant.* **2016**, *50*, 2591–2605. [CrossRef]

22. Bluszcz, A. European economies in terms of energy dependence. *Qual. Quant.* **2017**, *51*, 1531–1548. [CrossRef] [PubMed]

23. Kowal, B. Key performance indicators in a multi-dimensional performance card in the energy sector. *IOP Conf. Ser. Earth Environ. Sci.* **2019**, *214*, 012093. [CrossRef]

24. Rybak, A.; Rybak, A. Analysis of the Strategy for the Energy Policy of Poland until 2030 implementation effects in the aspect of environmental protection taking into account the energy security of Poland. *IOP Conf. Ser. Earth Environ. Sci.* **2019**, *261*, 012044. [CrossRef]

25. Rybak, A.; Rybak, A. Possible strategies for hard coal mining in Poland as a result of production function analysis. *Resour. Policy* **2016**, *50*, 27–33. [CrossRef]

26. Manowska, A. Renewable energy sources in the Polish energy structure. In *Bezpieczeństwo energetyczne—Filary i perspektywa rozwoju. Księga abstraktów IV konferencji naukowej, Rzeszów, 1-2 kwietnia 2019 r. Red*; Borkowski, P., Węgrzyn, P., Guźla, E., Siwiec, D., Czerwińska, K., Eds.; Instytut Polityki Energetycznej im. I. Łukasiewicza: Rzeszów, Poland, 2019; p. 27.

27. US Department of Energy. *Levelized Cost of Energy (LCOE)*; Annual Energy Outlook 2019; U.S. Energy Information Administration: Washington, DC, USA, 2019; pp. 1–9.

28. Luque, A.; Hegedus, S. *Handbook of Photovoltaic Science and Engineering*; Wiley & Sons: Hoboken, NJ, USA, 2010; p. 235.

Article

Indirect Convective Solar Drying Process of Pineapples as Part of Circular Economy Strategy

Yaovi Ouézou Azouma [1], Lynn Drigalski [2], Zdeněk Jegla [3,*], Marcus Reppich [2], Vojtěch Turek [3] and Maximilian Weiß [2]

[1] Ecole Supérieure d'Agronomie, Université de Lomé, Lomé BP 1515, Togo
[2] Faculty of Mechanical and Process Engineering, Augsburg University of Applied Sciences, An der Hochschule 1, 86161 Augsburg, Germany
[3] Faculty of Mechanical Engineering, Institute of Process Engineering, Brno University of Technology, Technická 2, 61669 Brno, Czech Republic
*   Correspondence: zdenek.jegla@vut.cz

Received: 28 June 2019; Accepted: 20 July 2019; Published: 24 July 2019

Abstract: This study investigates the industrial-scale application of a simple convective solar drying process of pineapples as part of a circular economy strategy for developing countries. A renewable energy concept is presented, which follows the circular economy aims by effectively employing a simple system for biogas production and a two-stage drying system. Both these systems meet the requirements for implementation in the specific conditions of developing countries, of which Togo, where pineapple is a major crop, is taken as an example. With respect to earlier findings available in the literature, the paper focuses on the solar drying process, which is critical to the proposed strategy. A portable solar dryer working in indirect heating mode was built and later also modified to enhance its performance. Three main factors influencing the convective drying process, namely, drying time (270 min, 480 min), solar radiation intensity (650 W/m^2, 1100 W/m^2), and slice thickness (6–8 mm, 12–14 mm), were considered. The statistical Design of Experiments (DOE) method was applied to reduce the number and scope of experiments. In the best case, the moisture content was reduced from 87.3 wt % in fresh samples to 29.4 wt % in dried samples, which did not meet the quality requirements for dried fruit. An additional conventional post-solar drying procedure would, therefore, still be necessary. Nonetheless, the results show that in the case of pineapple drying the consumption of fossil fuels can be decreased significantly if convective solar pre-drying is employed.

Keywords: drying; solar energy; sustainable processing; energy efficiency

---

1. Introduction

Sustainable food processing is becoming an increasingly important issue in developing countries. To improve the local living conditions and redress global inequalities, a system-oriented approach in food production considering the whole value chain, including the economic, environmental, and social impacts, is essential. Moreover, future economic development depends particularly on how processing capacities in the agricultural sector can successfully be improved instead of exporting agricultural products unprocessed, and with no value added, to foreign markets. In this way, new quality jobs can be created at the same time.

Pineapple, after banana and citrus, is the third most important tropical fruit in international trade [1]. The seasonal pineapple processing in Togo, where pineapple is a major crop, was selected for the present study. According to [2], in 2016, Togolese farmers grew pineapples on a cultivation area of 237 ha and the annual production was 1908 t. The main commercial pineapple products are currently canned pineapple slices and chunks, juice and nectar, and frozen and dried fruit. Of these, the most common are dried fruit and juice, usually in organic quality, which are exported primarily to

European countries. The processing itself takes place in various decentralized small- and medium-sized enterprises. As described in [3], a typical such enterprise processes 1–2 t/d of fresh fruit, while the resulting pineapple waste (peels, cores, stems, and crowns) is about 40% of this amount. Other sources give even higher figures up to 75% [4]. The waste, together with discarded fruits with high moisture content and various other farm production wastes (cow dung, etc.) can be efficiently utilized for energy recovery via biogas production, and the resulting sludge can be used as fertilizer. While sophisticated biogas production technologies are employed in the developed countries, the technologies in developing countries must be tailored to the local conditions. Both the adaptation of European biogas production technologies to the specific requirements of pineapple waste processing in Togo and the corresponding experimental results were discussed by the authors of the present paper in [3].

1.1. Key Role of the Drying Process

Drying—that is, a continuous or intermittent process associated with heat and mass transfer—significantly influences the shelf life, appearance, composition, taste, shape, structure, and other characteristics of the product. With respect to the fact that this process is probably the most energy-intensive one in the food processing industry [5], any small increase in energy efficiency will contribute to sustainable development in the respective industrial sector. It is obvious that moisture content in fresh fruit (generally about 85 wt % [1]) is of crucial importance. Water activity, that is, the ratio of vapor pressure to saturation vapor pressure at a given temperature, must also be considered, because it too affects the shelf life of a food product [6]. This ratio should be around 20% to reduce physical, chemical, and biochemical reactions and to minimize microbiological growth in a food product [7].

Recent developments in drying technologies of agricultural produce in general were discussed in [8], while another study [9] focused solely on fruits and vegetables. Comprehensive reviews of various drying technologies used in Africa [10] or just the sub-Saharan zone [11], but without any special focus on the feasibility of (semi-)industrial-scale indirect solar drying of pineapples, are also available. Similarly, reviews focusing on the drying of single fruit (e.g., fig [12], mango [13], or mulberry [14]) or vegetable varieties (e.g., chili peppers [15]) have been published as well.

Hot air dryers have been addressed abundantly in the literature, mainly with respect to the properties of the dried produce. One can find studies discussing color change [16], shrinkage [17], remaining bioactive compounds [18], microstructure [19], effective moisture diffusivity [20], surface polyphenol accumulation [21], or even shear strength [22]. Apart from those mentioned above, there are also papers focusing on the optimum hot air dryer setup [23], real-time monitoring of the properties of the product being dried [24], the effects of pre-treatments [25], low-temperature drying [26], or various combined drying processes such as, for example, freezing-hot air drying [27], ultrasound-assisted hot air drying [28], or pressure drop-assisted hot air drying [29].

Considering the solar drying process in particular, one can encounter review articles covering both solar drying in specific African countries [30] and regions outside of Africa [31]. Analogously, there are papers focusing on the solar drying of banana [32], date [33], apple [34], coconut [35], Moroccan sweet cherry [36], grapes [37], blackberry [38], cocoa beans [39], tomato [40], potato [41], black turmeric [42], lemon balm leaves [43], medicinal plants [44], ber fruit [45], or algae [46]. Studies discussing in detail the solar drying of pineapple, however, are relatively rare, and they involve prohibitively long drying times [47] (which are not really feasible if the process is to be implemented (semi-)industrially), focus more on product shelf life and sensorial and bromatological analysis [48] than on the actual drying process, discuss a specific dryer chamber design in terms of a Computational Fluid Dynamics (CFD) simulation of air flow therein [49], or address drying of thin pineapple slices pre-treated with a sucrose solution of various concentrations [50] or coated with different edible coatings [51]. Other studies focus on drying in a ventilated tunnel [52] instead of in a drying chamber into which hot air is supplied from a solar thermal collector, or employ various mixed mode [53] or hybrid [54] dryers, where both direct and indirect solar radiation is in effect or different hot media are used instead of air, respectively.

1.2. Circular Economy Strategy for Pineapple Processing in Togo

Togolese pineapple processing companies currently use conventional dryers, where the heat required for drying is generated by the combustion of liquefied gases (e.g., propane or butane) [55]. The fuel is bought in gas cylinders and thus the process is very demanding in terms of operating cost. Moreover, the liquefied gas supply is not sufficiently decentralized, and, therefore, at least a partial shift to renewable energy sources would be desirable. Following the findings presented in [3], the current paper proposes a circular economy strategy for pineapple processing in Togo, which is shown in Figure 1. This strategy, being somewhat similar to the one presented in [56] which involves cattle market wastes in Nigeria, lies in the efficient utilization of pineapple processing wastes. The starting point is fresh pineapple (as highlighted in Figure 1), from which two cycles are originating—the production cycle and the waste cycle. The production cycle involves the preparation of slices, which are then pre-dried in the solar dryer and, as an intermediate product, continue to the conventional dryer. There, the final product (dried pineapple fruit) is obtained. The waste cycle begins with the wastes, which are inoculated with cow dung and enter the digester, where biogas is produced. This gas, being the primary product of the cycle, is then used as a fuel in the conventional dryer. The secondary waste cycle product (sludge) is utilized as fertilizer in pineapple cultivation, and thus the circular economy cycle is closed.

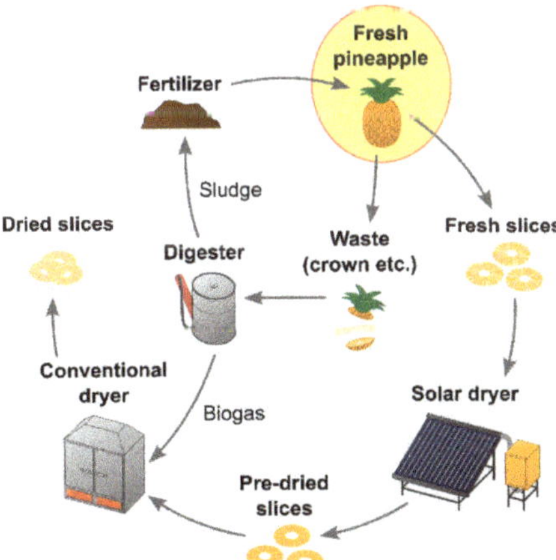

Figure 1. The circular economy principle presented for the example of pineapple processing in Togo.

For the strategy in Figure 1 to fully work, the biogas production (as discussed in [3]) must cover the fuel consumption in the conventional dryer. Therefore, the amount of fuel needed is decreased by pre-drying the pineapple slices in the solar dryer and, consequently, operating parameters of the dryer are essential for the optimal setup of the entire system. This is why the remaining portion of the present paper discusses in detail the solar drying process and the results of the corresponding pineapple drying experiments. The study's aim was to establish whether the industrial-scale application of an indirect convective solar dryer, implemented as indicated in Figure 1, would be feasible in the West African region. Further objectives were to determine the main parameters influencing the drying process and their optimal values, and to quantify the resulting decrease in the consumption of fossil fuels. The discussed energy-saving drying process, which preserves the nutritional quality of dried pineapple fruit, would then provide farmers with limited access to fossil fuels an option to process their agricultural products locally in a simple yet reliable manner.

2. Materials and Methods

For this initial feasibility study, a laboratory-scale solar dryer for pineapple processing was built and tested. The regional conditions (i.e., the temperature and solar radiation availability in Togo) were reproduced using an indoor test facility for solar thermal collectors and photovoltaic panels. A two-level factorial design was used to evaluate the main factors affecting the drying process and their interactions. The experimental work was conducted at Augsburg University of Applied Sciences, Laboratory of Energy and Process Engineering, Germany.

2.1. Experimental Device and Measuring Devices

The newly built experimental device used was a convective solar dryer consisting of a 1.5×0.8 m solar thermal collector and a drying chamber with the dimensions of $0.47 \times 0.37 \times 0.6$ m. The non-concentrating solar thermal collector for low-thermal applications adopted a modified flat plate format with a zig-zag geometry of the aluminum absorber, resulting in a surface area of 1.49 m^2 (Figure 2a). The bottom of the collector was insulated using a 20 mm thick expanded polystyrene panel (EPS; thermal conductivity: 0.034 W/(m K)). The wooden drying chamber was equipped with three drying trays with wire meshing, each having a net surface area of 0.11 m^2 (Figure 2b). The collector and the drying chamber were connected using a flexible aluminum tube insulated using a 25 mm thick, foil-faced polyethylene foam layer (thermal conductivity: 0.400 W/(m K), see Figure 3). The air flow around the material to be dried was ensured by the presence of a 2 W axial fan (type EE92251S1-000U-A99, Sunonwealth Electric Machine Industry Co. Ltd., Kaohsiung, Taiwan; flow rate up to 87.4 m^3/h) at the solar collector outlet. Power was supplied to the fan by a small photovoltaic module with a capacity of 5 W$_p$. The experimental device was designed using inexpensive materials which are readily available in Togo. The total cost amounted to 377 EUR [57].

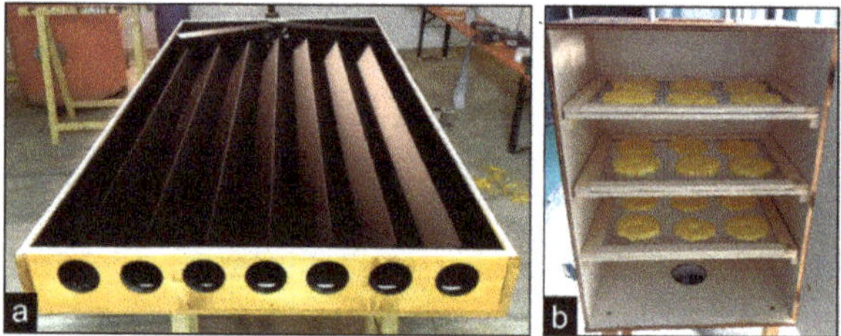

Figure 2. Main parts of the solar dryer: (**a**) the experimental solar collector (without collector cover) and (**b**) the drying chamber.

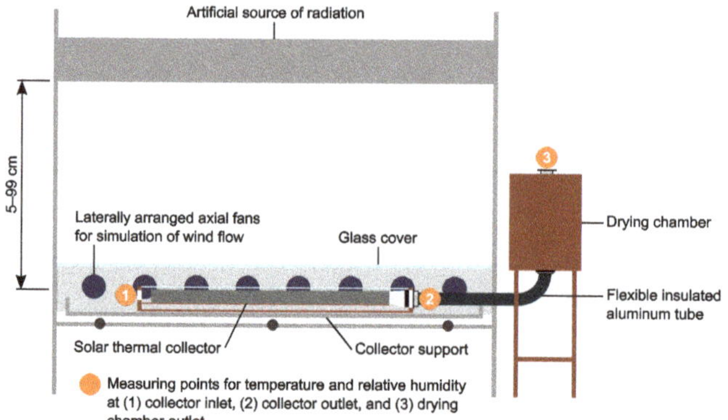

Figure 3. Indoor test facility with the solar collector, drying chamber, and relevant measuring points: (**1**) solar collector inlet, (**2**) solar collector outlet, and (**3**) drying chamber outlet.

The working principle of the solar drying system is based on the conversion of solar radiation to thermal energy in the black-coated aluminum absorber shown in Figure 2a. The absorber itself features a high absorption factor of more than 90% and a high thermal conductivity of 215 W/(m K). A transparent acrylic glass cover with high transmittance of 90% and high mechanical resistance reduces thermal losses to the surroundings. Simultaneously, the acrylic glass cover lets through only very little long-wave radiation emitted by the absorber. The air that acts as the heat transfer medium flows through and is heated up in the space between the absorber and the cover. The heated air is then passed through the drying chamber, where it circulates around the dried material by means of forced convection. This results in a gentle drying process without direct contact of the product with solar radiation.

The solar thermal collector was placed in the indoor test facility (Figure 3) containing 28 mercury vapor lamps and 27 halogen lamps. The distance between the source of radiation and the collector was variable between 0.05 m and 0.99 m so that different radiation intensities could be simulated. The respective drying experiments were carried out with two different solar radiation intensities, 650 W/m^2 and 1100 W/m^2. Eight laterally arranged fans were used for wind simulation. The test unit was equipped with Testo 635 sensors (Testo SE & Co. KGaA, Titisee-Neustadt, Germany) to measure temperature (range: −20 °C to +70 °C, resolution: 0.1 °C, accuracy: ±0.3 K) and relative humidity (0–100%, 0.1%, ±2%) of the air at the inlet and outlet of the solar collector and at the drying chamber outlet as shown in Figure 3. It was assumed that the temperature and relative humidity at the solar collector outlet were approximately equivalent to the conditions at the drying chamber inlet. A pyranometer type 8101/8102 (Philipp Schenk GmbH, Vienna, Austria) was used to measure solar radiation (measuring range: 0–1500 W/m^2, spectral range: 0.3–3 µm, resolution: 1 W/m^2, accuracy: ±3%). To obtain the air volumetric flow rate, a hot wire anemometer type FV A915 S120 (Ahlborn Mess- und Regelungstechnik GmbH, Holzkirchen, Germany) was used (measuring range: 0.1–25 m/s, resolution: 0.01 m/s, accuracy: ±5%). Weight of the samples was measured using a Mettler PM 4600 electronic scale (N.V. Mettler-Toledo S.A., Zaventem, Belgium; accuracy: ±0.02 g).

2.2. Preparation of Samples

Pineapples were peeled, cored, trimmed, and cut into slices in a single step using a pineapple slicer (Figure 4). The diameter of the prepared slices was 90–100 mm. In order to identify the effects of slice thickness on drying, two varying thicknesses, 6–8 mm and 12–14 mm, were considered. In addition, the rings were cut into eighths in order to examine the drying kinetics. The initial moisture content (87.3 ± 1.2 wt %) was determined according to DIN EN 322 [58] and DIN EN 15414-3 [59], that is, the samples were weighed before being placed in an oven at 105 ± 2 °C for 24 h to be fully dried, and then they were weighed again.

Figure 4. Preparation of samples: (**1**) fresh pineapples, (**2**) cutting of leaf crown, (**3**) cutting by means of rotational motion, (**4**) removing of the flesh, and (**5**) final spiral of flesh without core.

Commercially produced dried pineapple fruit served as a reference product and its moisture content (13.7 ± 0.67 wt %) was obtained in the same manner. This was then used to evaluate the performance of the solar dryer.

2.3. Experimental Parameters

To reduce the number and the scope of experiments and to discover the relationships between the factors affecting the drying process, the statistical Design of Experiments (DOE) method was applied [60]. The most significant factors influencing the drying process identified as controllable input factors were the drying time, solar radiation intensity, and slice thickness. The effect of air flow rate was not investigated in this preliminary feasibility study, but was selected according to the available axial fan performance and cost, fan characteristic curve, and the total pressure drop in the solar drying system (42.1 m^3/h, corresponds to roughly half the maximum fan throughput). Other controllable input factors such as mean wind velocity of 1.8 m/s, the angle of incidence of solar radiation of 90°, and the diameter of pineapple slices were always the same as well. Uncontrollable input factors were temperature and relative humidity of the ambient air and the moisture content in the fresh pineapple. The moisture content in the dried pineapple was the output used for validating the solar drying process. The controllable input factors were varied in this experimental design at two levels, low (−) and high (+), as shown in Table 1.

Table 1. Two-level factorial design of controllable input factors and their associated levels.

Factor	Name	Low Level (−)	High Level (+)
A	Drying time	270 min	480 min
B	Solar radiation intensity	650 W/m^2	1100 W/m^2
C	Slice thickness	6–8 mm	12–14 mm

As for the drying time, this was varied with respect to the minimum and maximum usable daily sunshine duration in Togo in the months of June to October and November to May (the primary pineapple processing periods in this country). The selected solar radiation intensities corresponded to the respective prevailing minimum and maximum values. All the relevant climate data were taken from the software Meteonorm 7 (Meteotest AG, Bern, Switzerland) [61]. A full factorial design with k factors attaining two levels was chosen [62]. The number of experiments was therefore given by $n = 2^k$, that is, for $k = 3$ the experimental design comprised eight tests as shown in Table 2. This experimental design was carried out in a random order twice to determine the influence of input factors on the output more accurately and to mitigate the effect of scattering.

Table 2. Setting of input factors according to the full factorial design.

Test Number	Factor A	Factor B	Factor C
1	−	−	−
2	+	−	−
3	−	+	−
4	+	+	−
5	−	−	+
6	+	−	+
7	−	+	+
8	+	+	+

The moisture content in the dried pineapple was not influenced by only the individual input factors, A, B, and C. Therefore, the effects of all possible interactions of two factors, and of all three factors mentioned above, also had to be investigated using analysis of variance. For an experiment design comprising 2^k experiments, $2^k - 1$ effects could be identified with positive or negative signs as presented in Table 3. The signs of individual input factors were equal to their levels from Table 2. In each row, the signs of interactions, AB, AC, BC, and ABC, correspond to the products of signs of the respective input factors. The effects can then be estimated as follows:

$$\text{effect} = \frac{2}{n} \sum_{i=1}^{n} (\text{sign} \cdot \bar{y}_i), \tag{1}$$

where i denotes the test number and $\bar{y}_i$ is the mean of the respective experimental results.

Table 3. Design matrix and signs for seven effects in the 2^3 full factorial design.

Test Number	Factors			Interactions			
	A	B	C	AB	AC	BC	ABC
1	−	−	−	+	+	+	−
2	+	−	−	−	−	+	+
3	−	+	−	−	+	−	+
4	+	+	−	+	−	−	−
5	−	−	+	+	−	−	+
6	+	−	+	−	+	−	−
7	−	+	+	−	−	+	−
8	+	+	+	+	+	+	+

2.4. Drying Process

In a typical medium-sized Togolese company, around 1.5 t/d of fresh pineapples are processed, of which roughly 900 kg/d are dried at 50–60 °C for 20 h. The remaining 600 kg/d leave the production process as waste. For the conventional drying process, the daily butane consumption is around 49.5 kg which corresponds to a daily cost of 39.20 EUR [55]. Assuming a calorific value of butane of 12.72 kWh/kg at 25 °C, the daily energy consumption required for drying is 629.6 kWh or 0.7 kWh per kilogram of fresh fruit.

For the experiments, the indoor test facility was switched on and the drying chamber was preheated without the trays for 15 min. Then, six prepared pineapple slices were placed at different locations on each tray, and the trays were placed into the drying chamber. Subsequently, continuous drying tests were run with the parameters being set according to Table 2. The temperature and relative humidity of air were measured and recorded every two minutes during the entire experimental period. Afterwards, the pineapple slices were crushed, and their moisture content was determined. The moisture content in the product to be dried was expressed on total material basis as $W = m_W / (m_{DM} + m_W)$, where m_W is the mass of water contained in the pineapples and m_{DM} is the respective dry solid mass. The moisture content can also be expressed on the dry basis,

$$X = \frac{m_W}{m_{DM}}, \tag{2}$$

where the mass of the dry solid, m_{DM}, remains constant during the entire drying process. Both values of product moisture can be converted between each other using $X = W/(W-1)$.

The ambient air entering the solar collector was characterized by the temperature of 25 °C, relative humidity of 30 wt %, and pressure of 1.012 bar. The simplified equation,

$$\dot{m}_{air} = \frac{\Delta \dot{m}_W}{x_{out} - x_{in}}, \tag{3}$$

assuming a continuous, steady-state dryer operation, can be used for the calculation of drying air consumption. Here, the moisture removed from a product to be dried is defined as $\Delta \dot{m}_W = \dot{m}_{V,out} - \dot{m}_{V,in}$ and the absolute air humidity is defined as $x = m_V / m_A$. In these equations, $\dot{m}_{V,in}$ and $\dot{m}_{V,out}$ denote the vapor inlet and outlet mass flow rates, respectively, m_V is the mass of vapor, and m_A is the mass of dry air. The enthalpy of moist air is then given by the following equation:

$$h = h_A + x h_V = c_{p,A} t + x \left(L + c_{p,V} t \right), \tag{4}$$

where h_A denotes the enthalpy of dry air, h_V is the enthalpy of vapor, $c_{p,A}$ and $c_{p,V}$ are the specific heat capacities of dry air and vapor, respectively, t is the temperature, and L is the specific heat of vaporization. The corresponding absolute humidity can be calculated using the following equation:

$$x = \frac{R_A}{R_V} \left(\frac{\varphi p_S}{p - \varphi p_S} \right), \tag{5}$$

where φ denotes the relative air humidity, $R_A = 0.2871$ kJ/(mol K) and $R_V = 0.4614$ kJ/(mol K) are the specific gas constants of dry air and vapor, respectively, p_S is the saturation pressure, and p is the actual pressure. The energy required by the drying process is then calculated as $\dot{Q}_D = \dot{m}_{air} (h_{in} - h_{out}) = \dot{m}_{air} \cdot h_D$, where h_{in} and h_{out} denote the enthalpies of air at drying chamber inlet and outlet, respectively, and Δh_D is the change in air enthalpy in the drying chamber.

The source of heat in the solar thermal collector was the radiation generated by the lamps in the indoor test facility. The input power is usually the solar radiation received by the surface of the collector, absorbed and transferred to the drying air. This must be heated from ambient conditions at around 25 °C to conditions required at the inlet of the drying chamber (at least 55 °C). The heat flux to the drying air is therefore given by the following equation:

$$\dot{Q}_C = \dot{m}_{air} (h_{in} - h_{amb}) = \dot{m}_{air} \cdot h_C, \tag{6}$$

where h_{amb} denotes the enthalpy of ambient air and Δh_C is the change in air enthalpy in the solar thermal collector. The same heat flux can be written in terms of quantities representing the energy irradiated to the thermal collector and the losses, $\dot{Q}_{loss}$, as $\dot{Q}_C = \eta_0 E A_C - \dot{Q}_{loss}$, where η_0 denotes the optical efficiency of the solar thermal collector and A_C is its area. The overall efficiency of the solar thermal collector, including optical and thermal losses, can then be obtained using the following equation:

$$\eta_C = \eta_0 - \frac{a(t_C - t_{amb})}{E}, \tag{7}$$

where η_0 and a are constants to be evaluated either analytically or experimentally, while t_C denotes the mean collector temperature and t_{amb} is the ambient temperature. According to [63], the typical experimental coefficients in efficiency correlations for air collectors operating between 20 °C and 50 °C are $\eta_0 = (0.75–0.80)$ and $a = (8–30)$ W/(m^2 K). The required collector area then follows from the general definition of the overall efficiency of the solar thermal collector, that is,

$$A_C = \frac{\dot{Q}_C}{\eta_C E}. \tag{8}$$

3. Results and Discussion

The overall performance and efficiency of the solar collector used to provide hot air to the drying process were considered first. With respect to the proposed experimental design, the residual moisture of dried pineapple fruit was used as the metric. The drying kinetics were determined for both slice thicknesses. Finally, the impact of local climatic conditions on the drying process performance was estimated together with the expected fossil fuel savings, and the design of the solar thermal collector was improved.

3.1. Solar Thermal Collector Performance

Initially, the solar thermal collector was designed for the worst-case scenario, that is, the required drying time of 4 h and low radiation intensity of 650 W/m^2, obtained via Equations (3) and (5) for a minimum air flow rate of 33.4 m^3/h. Under these assumptions, the collector heat flux calculated using Equations (4)–(6) was 280.12 W. The necessary solar thermal collector area, A_C, was then determined using Equations (7) and (8) to be 1.49 m^2 for $\eta_0 = 0.75$, $a = 20$ W/(m^2 K), and ($t_C - t_{amb}$) = 15 K.

To determine the overall performance and efficiency of the solar thermal collector for each test, temperature and relative humidity of air at the collector inlet and outlet were recorded. In total, 16 tests at radiation intensities of 650 W/m^2 and 1100 W/m^2 were carried out. Figure 5 shows an example of typical measured values at the high solar radiation intensity of 1100 W/m^2 over the period of 480 min. For further consideration, the respective average temperatures and relative humidities after the warm-up phase of 50 min were used.

Table 4 summarizes for each test the increases in drying air temperature, $t_{in} - t_{amb}$. Using the corresponding mean value, the air temperature at the solar collector outlet was 46.8 °C at 650 W/m^2, whereas at 1100 W/m^2 it was 56.8 °C. This means that the required temperature of about 55 °C at the drying chamber inlet can only be guaranteed at the high solar radiation intensity level.

Table 4. Air temperature changes, $t_{in} - t_{amb}$ (K), between the solar collector inlet and outlet.

E, W/m^2	Test Number								
	1	2	3	4	5	6	7	8	Mean
650	20.81	20.66	20.49	20.48	19.90	21.38	20.50	20.04	20.53
1100	29.78	28.84	29.43	29.15	29.71	29.26	30.23	28.51	29.36

Given the air flow rate of 42.1 m^3/h and mean air density of 1.12 kg/m^3, the overall efficiency of the solar thermal collector was calculated using Equations (4)–(6) and (8). The results are listed in Table 5.

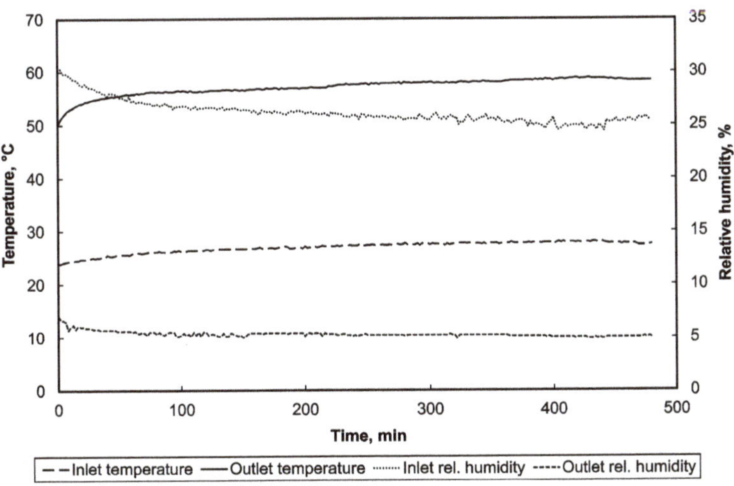

Figure 5. Typical dependence of air temperatures at the inlet and outlet of the solar thermal collector, and the corresponding air relative humidities, on drying time at the solar radiation intensity of 1100 W/m^2.

Table 5. Overall performance and efficiency of the solar thermal collector.

E, W/m^2	Δh_C, kJ/kg	$\dot{Q}_C$, W	η_C, –
650	20.90	269.58	0.298
1100	30.85	391.00	0.255

3.2. Results of the Experiments

The experiments were compared on the basis of moisture content in the pineapple fruit before and after the drying process. The eight tests of full factorial design as given in Tables 1 and 2 carried out twice yielded 16 moisture content values shown in Table 6. The data indicate that the best drying performance was reached in test number 4 with the drying time of 480 min, radiation intensity of 1100 W/m^2, and slice thickness of 6–8 mm (Figure 6a). Significant shrinkage was apparent in the case of the corresponding dried pineapple slice. The second-best result was obtained in test number 2. In contrast, test number 5 with the drying time of 270 min, solar radiation intensity of 650 W/m^2, and slice thickness of 12–14 mm showed the least satisfactory outcome with the highest residual moisture content (Figure 6b). Compared to a fresh pineapple slice, here there was only a slight change in appearance due to the removal of moisture just from the surface.

Table 6. Residual relative moisture content (wt %) for the eight tests comprising the full factorial design.

Test Number	Test Run 1	Test Run 2	Mean	Variance	Standard Deviation
1	63.48	68.52	66.00	12.70	3.56
2	49.18	47.51	48.35	1.40	1.18
3	66.11	55.15	60.63	60.06	7.75
4	26.33	32.52	29.42	19.15	4.38
5	78.69	78.08	78.38	0.19	0.43
6	69.56	71.79	70.68	2.49	1.58
7	74.49	77.10	75.80	3.41	1.85
8	67.43	62.99	65.21	9.82	3.13

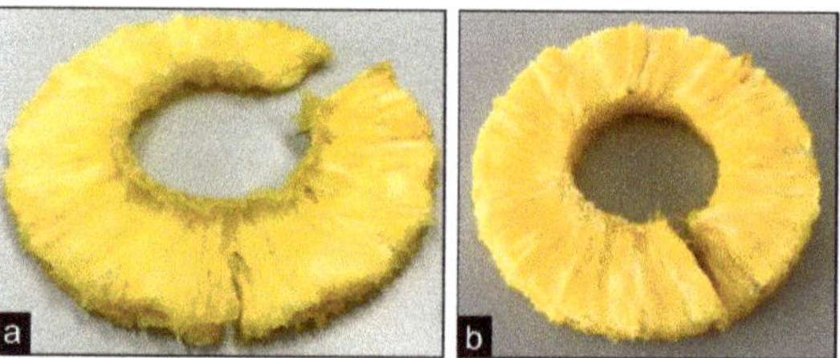

Figure 6. Visual comparison of the dried pineapple slices obtained using (**a**) the best combination of input factors and (**b**) the worst combination of input factors.

Using the factor signs from Table 3 and the means in Table 6, the means for the two levels, (−) and (+), can be calculated by column for each input factor and each combination of input factors. The results are shown in Figure 7. The greater the deviation between the two means, (−) and (+) (i.e., the steeper the line connecting both means), the greater the influence of a factor or a factor interaction on the drying process. It is obvious from Figure 7 that the moisture content in the dried pineapple fruit is most affected by slice thickness (factor C) and drying time (factor A). In comparison, solar radiation intensity (factor B), as well as all interactions of individual factors, influence the process output significantly less. Furthermore, the effects of the individual input factors as well as of the combinations of factors can be expressed quantitatively using Equation (1), as shown in Table 7.

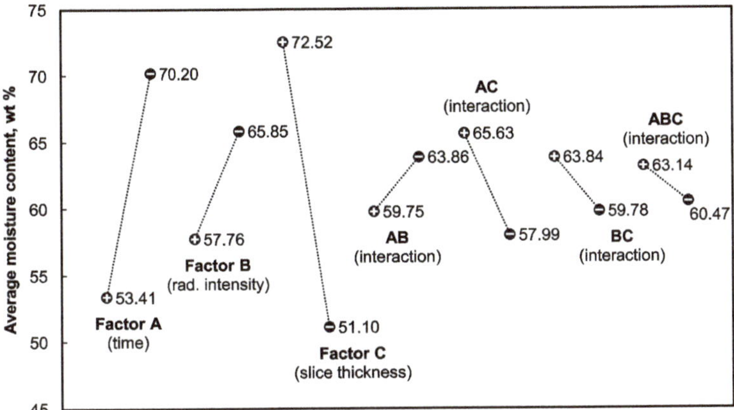

Figure 7. Effects of individual input factors, A, B, and C, and their interactions, AB, AC, BC, and ABC.

Table 7. Estimates of factor effects (wt %).

A	B	C	AB	AC	BC	ABC
16.79	8.09	−21.42	4.11	−7.64	−4.06	−2.67

The widths of the overall 95%, 99%, and 99.9% confidence intervals obtained using the data in Table 6 and other relevant values were ±4.26 wt %, ±6.20 wt %, and ±9.31 wt %, respectively. These are plotted in Figure 8 against the corresponding factor effects from Table 7. It is obvious that factors A and C are statistically highly significant, factor B and interaction AC are statistically significant, and the remaining interactions are statistically insignificant.

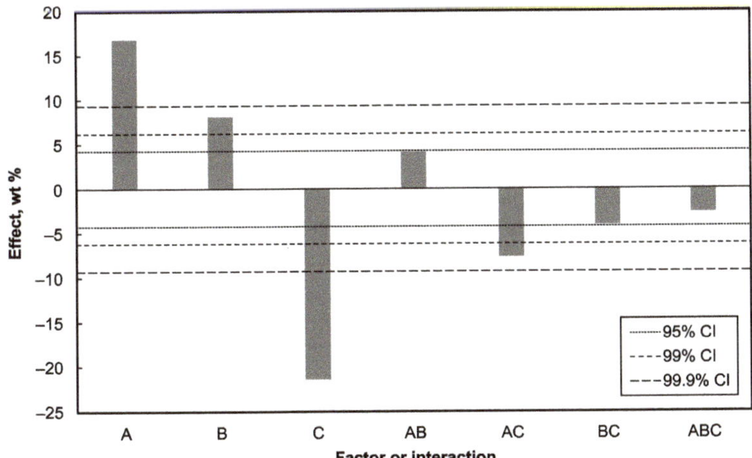

Figure 8. The overall 95%, 99%, and 99.9% confidence intervals (CI) plotted against the effects of the individual factors, A, B, and C, and their interactions, AB, AC, BC, and ABC.

The conclusions of the drying experiments are that the slice thickness of prepared fresh pineapples should not exceed 8 mm. The available daily sunshine duration limited by the climatic conditions should be utilized as much as possible. The solar drying process is affected by the variation of the annual solar radiation intensity to a lesser degree and, therefore, the solar thermal dryer can be used in the respective locality over the entire year. Performance of the drying process, however, also depends on the local climatic conditions, particularly on the temperature and humidity of the ambient air. In the case of Togo, the performance was verified as shown in Table 8. Table 9 then lists for Togo the air enthalpies, h_{in}, from Equation (6), temperatures, t_{in}, from Equation (4), and absolute humidities, x_{in}, from Equation (5) at the solar collector outlet corresponding to the two solar radiation intensity levels (i.e., the constant collector outputs of 270 W and 391 W).

In any case, the best drying result, characterized by a product moisture content of 29.4 wt %, still does not meet the quality of the reference product having the residual moisture content of 13.7 wt %. The solar thermal drying process must, therefore, be extended for a further 10–12 h period or another suitable post-solar drying procedure must be implemented.

Table 8. Climatic conditions in the laboratory and in Togo (average).

Parameter	Laboratory	Togo (Average)
Ambient air temperature	27 °C	28 °C
Relative humidity	0.27	0.85
Absolute humidity	5.89 g/kg	20.43 g/kg
Air pressure	1012 mbar	1008 mbar
Saturation vapor pressure	35.57 mbar	37.71 mbar
Ambient air enthalpy	42.03 kJ/kg	79.48 kJ/kg

Table 9. Calculated data for the average climatic conditions in Togo.

Parameter	$E = 650$ W/m^2	$E = 1100$ W/m^2
Solar collector outlet enthalpy	99.85 kJ/kg	109.02 kJ/kg
Solar collector outlet temperature	49.11 °C	57.86 °C
Relative humidity	0.27	0.18
Absolute humidity	20.43 g/kg	20.43 g/kg
Saturation vapor pressure	117.43 mbar	182.02 mbar

3.3. Drying Kinetics

The drying kinetics were determined by measuring the dry basis moisture content, X, according to Equation (2) and varying with time, for both pineapple slice thicknesses of 6–8 mm and 12–14 mm. The total drying time of 480 min was chosen together with the solar radiation intensity of 1100 W/m^2. Moisture content was measured every 30 min.

Figure 9 shows the drying kinetics for both slice thicknesses. The drying curve progressions can each be divided into two parts. In the case of the lower slice thickness of 6–8 mm (Figure 9a), the first part is characterized by a high constant drying rate, $dX/d\tau$, and the moisture content decreasing nearly linearly until the critical time $\tau_c = 350$ min is reached. Then, the drying process transitions from the external convective heat and mass transfer to the second part. where the drying rate is decreasing and is controlled by the internal diffusive mass transfer. For the slice thickness of 12–14 mm (Figure 9b), the drying curve is similar with the critical time being $\tau_c = 100$ min. In both cases, however, the lowest possible residual moisture content in the dried pineapple is limited by the equilibrium water content in the material at a given temperature characterized by the moisture sorption isotherm [64].

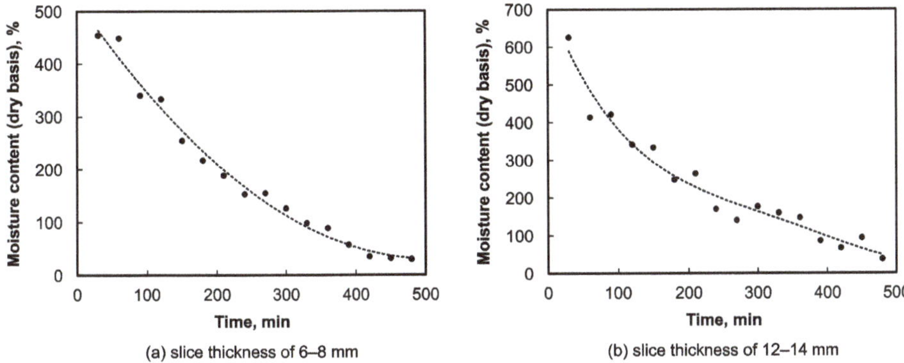

(a) slice thickness of 6–8 mm   (b) slice thickness of 12–14 mm

Figure 9. Drying kinetics of pineapple slices with the thickness of (**a**) 6–8 mm and (**b**) 12–14 mm.

3.4. Consumption of Fossil Fuels in the Post-Solar Drying Procedure

As stated in Section 2.4, for a typical pineapple processing rate of approximately 900 kg/d and conventional drying time of 20 h, the butane consumption is around 49.5 kg/d. Should the pineapple fruit be pre-dried using the solar dryer to the moisture content of 29.4 wt %, then the necessary post-solar drying time to reach the target moisture content of 13.7 wt % would be approximately 6.7 h. This translates to the butane consumption of 16.6 kg and, consequently, significant daily savings of 32.9 kg of butane, or 66%.

3.5. Modification of the Solar Collector Design

As can be seen from Table 9, nearly the same outlet temperatures of drying air are reached with both solar radiation intensities. However, higher absolute air humidity in Togo requires a higher air flow rate. The original solar thermal collector was therefore modified [65]. This entailed changes to the absorber and the thermal insulation to ensure a higher air temperature at the solar collector outlet as well as increased overall efficiency of the solar collector. The modified absorber was built from empty aluminum beverage cans with a commercial black acrylic varnish being applied (Figure 10). Mineral wool insulation (thermal conductivity: 0.040 W/(m K)) was used instead of expanded polystyrene due to higher air temperatures. The bottom insulation panel of the collector was twice as thick than before (40 mm) and the sides contained 10 mm thick insulation layers. In addition, the axial fan was equipped with a control system to regulate the outlet air temperature. The additional cost of all the modifications made amounted to 120 EUR.

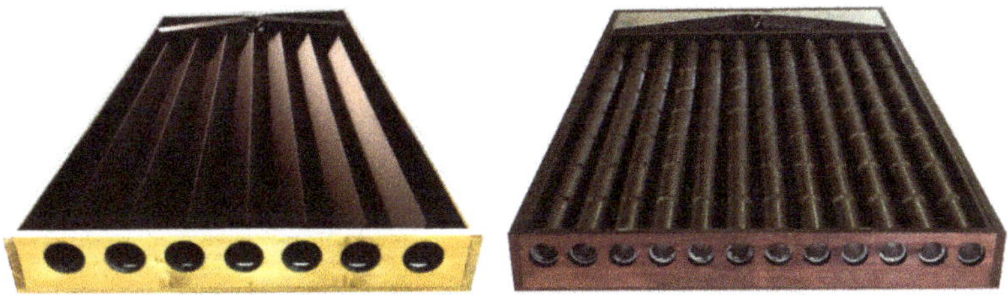

Figure 10. Comparison of the (**left**) original and (**right**) improved designs of the solar thermal collector.

The modified solar thermal collector was tested under the same conditions as mentioned in Section 3.1. With the solar radiation intensity of 1100 W/m^2, the resulting air temperature at the collector outlet was markedly higher (83.4 °C instead of the original 56.8 °C). The overall performance and efficiency of the modified solar thermal collector were also improved. The corresponding main results are summarized in Table 10.

Table 10. Main technical parameters of the modified solar thermal collector.

E, W/m^2	$t_{in} - t_{amb}$, K	t_{in}, °C	$\dot{Q}_C$, W	η_C, −
650	43.60	70.70	416.56	0.389
1100	54.10	83.40	884.94	0.489

4. Conclusions

This study investigated the feasibility of an industrial-scale application of indirect solar thermal drying of pineapples as part of a Togo-specific pineapple processing circular economy strategy, and the main factors affecting the respective drying process. Drying time and pineapple slice thickness proved to be significant, whereas the solar radiation intensity was found to affect the drying process to a lesser extent. As part of the experiments, drying kinetics were also determined for both slice thicknesses of 6–8 mm and 12–14 mm. The corresponding drying curves were characterized by two parts with different drying rates, which were delimited by the critical times of 350 min and 100 min, respectively.

Considering the best obtained residual moisture content in the dried pineapple fruit of 29.4 wt % and that of the reference product (13.7 wt %), the performance of the solar dryer alone was deemed insufficient. In order to meet the required product quality, either the solar thermal drying process had to be extended for a further 10–12 h period (which may not be feasible with respect to the local climatic conditions) or another post-solar drying procedure had to be added. Nevertheless, solar pre-drying of the pineapple fruit followed by a conventional drying process resulted in significant fossil fuel savings of around 66%. This lays the groundwork for the successful implementation of the proposed circular economy strategy.

A modification of the solar thermal collector to increase its efficiency was also briefly discussed. The improved design of the absorber, changes to the thermal insulation, and the addition of a fan control system resulted in a considerable increase in the drying air temperature at the collector outlet (83.4 °C instead of the original 56.8 °C).

Ultimately, the combination of the solar thermal drying process with post-solar drying based on the combustion of biogas produced from agricultural wastes leads to improved environmental sustainability and also supports the circular economy in the agricultural sector. The proposed strategy, therefore, could help farmers in developing countries to meet the growing demand for the sustainable processing of tropical fruit or other agricultural products.

Author Contributions: Conceptualization, M.R., Y.O.A., and Z.J.; methodology, M.R.; investigation, L.D. and M.W.; resources, M.R.; writing—original draft preparation, L.D., M.R., and M.W.; writing—review and editing, V.T. and Z.J.; visualization, V.T.

Funding: This research was funded by the Czech Ministry of Education, Youth and Sports/EU Operational Programme Research, Development and Education, Grant No. CZ.02.1.01/0.0/0.0/16_026/0008413, "Strategic partnership for environmental technologies and energy production", and by the Augsburg University of Applied Sciences.

Conflicts of Interest: The authors declare no conflict of interest.

Nomenclature

Roman Symbols:

A	surface area, m^2
a	constant in the collector efficiency correlation, –
c_p	specific heat capacity, kJ/(kg K)
E	solar radiation intensity, W/m^2
h	specific enthalpy, kJ/kg
k	number of controllable input factors, –
L	specific heat of vaporization, kJ/kg
m	mass, kg
$\dot{m}$	mass flow rate, kg/h
n	number of experiments, –
p	air pressure, Pa
p_s	saturation air pressure, Pa
$\dot{Q}$	heat flux, energy consumption, W
$\dot{Q}_{loss}$	energy losses from the solar thermal collector, W
R	specific gas constant, kJ/(mol K)
t	temperature, K
W	moisture content on total material basis, kg water/kg total matter, %
X	moisture content on dry matter basis, kg water/kg dry matter, %
x	absolute air humidity, kg water/kg dry air, –
$\bar{y}$	mean of the experimental results

Greek Symbols:

$\Delta\dot{m}_W$	moisture removed from pineapples, kg/h
τ	time, min
τ_c	critical time, min
η_0	optical efficiency of the solar collector, –
η_C	overall efficiency of the solar collector, –
φ	relative air humidity, –

Subscripts:

A	dry air
air	drying air
amb	ambient (at 25 °C, 30 wt % rel. humidity, 1.012 bar), kJ/kg
C	in/of the solar thermal collector
D	in the drying chamber, of the drying process
DM	dry matter
i	test number
in	at drying chamber inlet
out	at drying chamber outlet
V	water vapor
W	water

References

1. Lobo, M.G.; Paull, R.E. (Eds.) *Handbook of Pineapple Technology: Production, Postharvest Science, Processing and Nutrition*; John Wiley & Sons: Hoboken, NJ, USA, 2017.
2. FAOSTAT: Crops. Available online: http://www.fao.org/faostat/en/#data/QC (accessed on 25 June 2019).
3. Azouma, Y.O.; Jegla, Z.; Reppich, M.; Turek, V.; Weiß, M. Using agricultural waste for biogas production as a sustainable energy supply for developing countries. *Chem. Eng. Trans.* **2018**, *70*, 445–450. [CrossRef]
4. Roda, A.; De Faveri, D.M.; Giacosa, S.; Dordoni, R.; Lambri, M. Effect of pre-treatments on the saccharification of pineapple waste as a potential source for vinegar production. *J. Clean. Prod.* **2016**, *112*, 4477–4484. [CrossRef]
5. Kudra, T. Energy aspects in drying. *Dry. Technol.* **2004**, *22*, 917–932. [CrossRef]

6.   Staudt, P.B.; Kechinski, C.P.; Tessaro, I.C.; Marczak, L.D.F.; Soares, R.D.; Cardozo, N.S.M. A new method for predicting sorption isotherms at different temperatures using the BET model. *J. Food Eng.* **2013**, *114*, 139–145. [CrossRef]

7.   Kröll, K.; Kast, W. (Eds.) *Drying Technology, Vol. 3: Drying and Dryers in Production*; Springer: Berlin, Germany, 1989. (In German)

8.   Lamidi, R.O.; Jiang, L.; Pathare, P.B.; Wang, Y.D.; Roskilly, A.P. Recent advances in sustainable drying of agricultural produce: A review. *Appl. Energy* **2019**, *233*, 367–385. [CrossRef]

9.   Li, K.; Zhang, M.; Mujumdar, A.S.; Chitrakar, B. Recent developments in physical field-based drying techniques for fruits and vegetables. *Dry. Technol.* **2019**. [CrossRef]

10. Ndukwu, M.C.; Bennamoun, L.; Abam, F.I. Experience of solar drying in Africa: Presentation of designs, operations, and models. *Food Eng. Rev.* **2018**, *10*, 211–244. [CrossRef]

11. Boroze, T.; Desmorieux, H.; Méot, J.-M.; Marouzé, C.; Azouma, Y.; Napo, K. Inventory and comparative characteristics of dryers used in the sub-Saharan zone: Criteria influencing dryer choice. *Renew. Sustain. Energy Rev.* **2014**, *40*, 1240–1259. [CrossRef]

12. Mat Desa, W.N.; Mohammad, M.; Fudholi, A. Review of drying technology of Figure. *Trends Food Sci. Technol.* **2019**, *88*, 93–103. [CrossRef]

13. Cheng, W.; Sørensen, K.M.; Mongi, R.J.; Ndabikunze, B.K.; Chove, B.E.; Sun, D.-W.; Engelsen, S.B. A comparative study of mango solar drying methods by visible and near-infrared spectroscopy coupled with ANOVA-simultaneous component analysis (ASCA). *LWT* **2019**, *112*, 108214. [CrossRef]

14. Khattak, M.K.; Hanif, M.; Khan, M.; Ramzan, M.; Abdurab. Comparison of drying process and preservatives on drying kinetics, texture and antioxidants retention in mulberry fruits. *J. Anim. Plant Sci.* **2019**, *29*, 803–808.

15. Ayua, E.; Mugalavai, V.; Simon, J.; Weller, S.; Obura, P.; Nyabinda, N. Comparison of a mixed modes solar dryer to a direct mode solar dryer for African indigenous vegetable and chili processing. *J. Food Process. Preserv.* **2017**, *41*, e13216. [CrossRef]

16. Yang, X.-H.; Deng, L.-Z.; Mujumdar, A.S.; Xiao, H.-W.; Zhang, Q.; Kan, Z. Evolution and modeling of colour changes of red pepper (*Capsicum annuum* L.) during hot air drying. *J. Food Eng.* **2018**, *231*, 101–108. [CrossRef]

17. Nguyen, T.K.; Mondor, M.; Ratti, C. Shrinkage of cellular food during air drying. *J. Food Eng.* **2018**, *230*, 8–17. [CrossRef]

18. Süfer, Ö.; Palazoğlu, T.K. A study on hot-air drying of pomegranate. *J. Therm. Anal. Calorim.* **2019**. [CrossRef]

19. Talens, C.; Castro-Giraldez, M.; Fito, P.J. Effect of microwave power coupled with hot air drying on sorption isotherms and microstructure of orange peel. *Food Bioprocess Technol.* **2018**, *11*, 723–734. [CrossRef]

20. Sarobol, M.; Sarobol, P.; Teeta, S.; Pharanat, W. Investigation effective moisture diffusivity and activation Eenergy on convective hot air drying assisted extraction of dragon fruit slices. *J. Phys. Conf. Ser.* **2018**, *1144*, 012062. [CrossRef]

21. Li, X.; Wu, X.; Bi, J.; Liu, X.; Li, X.; Guo, C. Polyphenols accumulation effects on surface color variation in apple slices hot air drying process. *LWT* **2019**, *108*, 421–428. [CrossRef]

22. Elmizadeh, A.; Shahedi, M.; Hamdami, N. Quality assessment of electrohydrodynamic and hot-air drying of quince slice. *Ind. Crops Prod.* **2018**, *116*, 35–40. [CrossRef]

23. Zhu, A. The convective hot air drying of Lactuca sativa slices. *Int. J. Green Energy* **2018**, *15*, 201–207. [CrossRef]

24. Moscetti, R.; Raponi, F.; Ferri, S.; Colantoni, A.; Monarca, D.; Massantini, R. Real-time monitoring of organic apple (var. Gala) during hot-air drying using near-infrared spectroscopy. *J. Food Eng.* **2018**, *222*, 139–150. [CrossRef]

25. Önal, B.; Adiletta, G.; Crescitelli, A.; Di Matteo, M.; Russo, P. Optimization of hot air drying temperature combined with pre-treatment to improve physico-chemical and nutritional quality of 'Annurca' apple. *Food Bioprod. Process.* **2019**, *115*, 87–99. [CrossRef]

26. Červenka, L.; Červenková, Z.; Velichová, H. Is air-drying of plant-based food at low temperature really favorable? A meta-analytical approach to ascorbic acid, total phenolic, and total flavonoid contents. *Food Rev. Int.* **2018**, *34*, 434–446. [CrossRef]

27. Zielinska, M.; Ropelewska, E.; Zapotoczny, P. Effects of freezing and hot air drying on the physical, morphological and thermal properties of cranberries (Vaccinium macrocarpon). *Food Bioprod. Process.* **2018**, *110*, 40–49. [CrossRef]

28. Tao, Y.; Han, M.; Gao, X.; Han, Y.; Show, P.-L.; Liu, C.; Ye, X.; Xie, G. Applications of water blanching, surface contacting ultrasound-assisted air drying, and their combination for dehydration of white cabbage: Drying mechanism, bioactive profile, color and rehydration property. *Ultrason. Sonochem.* **2019**, *53*, 192–201. [CrossRef] [PubMed]

29. Zhou, L.; Wang, X.; Bi, J.; Liu, X.; Yi, J.; Wu, X. Effect of different moisture equilibration process on the quality of apple chips dried by instant controlled pressure drop (dic)-assisted hot air drying. *J. Food Process. Preserv.* **2018**, *42*, e13316. [CrossRef]

30. Kessy, R.F.; Ochieng, J.; Afari-Sefa, V.; Chagomoka, T.; Nenguwo, N. Solar-dried traditional African vegetables in rural Tanzania: Awareness, perceptions, and factors affecting purchase decisions. *Econ. Bot.* **2018**, *72*, 367–379. [CrossRef]

31. Fudholi, A.; Ridwan, A.; Yendra, R.; Desvina, A.P.; Hartono, H.; Majahar, M.K.; Suyono, T.; Sopian, K. Solar drying technology in Indonesia: An overview. *Int. J. Power Electron. Drive Syst.* **2018**, *9*, 1804–1813. [CrossRef]

32. Hatami, S.; Sadeghi, M.; Mireei, S.A. Indirect forced solar drying of banana slices: Phenomenological explanation of non-isotropic shrinkage and color changes kinetics. *Int. J. Green Energy* **2017**, *14*, 1277–1283. [CrossRef]

33. Seerangurayar, T.; Al-Ismaili, A.M.; Janitha Jeewantha, L.H.; Al-Nabhani, A. Experimental investigation of shrinkage and microstructural properties of date fruits at three solar drying methods. *Sol. Energy* **2019**, *180*, 445–455. [CrossRef]

34. Veeramanipriya, E. Drying kinetics of forced convection solar dryer for fruit drying. *Int. J. Recent Technol. Eng.* **2019**, *7*, 323–327.

35. Ayyappan, S. Performance and CO_2 mitigation analysis of a solar greenhouse dryer for coconut drying. *Energy Environ.* **2018**, *29*, 1482–1494. [CrossRef]

36. Ouaabou, R.; Nabil, B.; Hidar, N.; Lahnine, L.; Idlimam, A.; Lamharrar, A.; Hanine, H.; Mahrouz, M. Valorization of solar drying process in the production of dried Moroccan sweet cherries. *Sol. Energy* **2018**, *172*, 158–164. [CrossRef]

37. Hamdi, I.; Kooli, S.; Elkhadraoui, A.; Azaizia, Z.; Abdelhamid, F.; Guizani, A. Experimental study and numerical modeling for drying grapes under solar greenhouse. *Renew. Energy* **2018**, *127*, 936–946. [CrossRef]

38. López-Vidaña, E.C.; Figueroa, I.P.; Marcos, E.G.A.; Navarro-Ocaña, A.; Hernández-Vázquez, L.; Santiago-Urbina, J.A. Solar drying kinetics and bioactive compounds of blackberry (Rubus fruticosus). *J. Food Process Eng.* **2019**, *42*, e13018. [CrossRef]

39. Koua, B.K.; Koffi, P.M.E.; Gbaha, P. Evolution of shrinkage, real density, porosity, heat and mass transfer coefficients during indirect solar drying of cocoa beans. *J. Saudi Soc. Agric. Sci.* **2019**, *18*, 72–82. [CrossRef]

40. Fterich, M.; Chouikhi, H.; Bentaher, H.; Maalej, A. Experimental parametric study of a mixed-mode forced convection solar dryer equipped with a PV/T air collector. *Sol. Energy* **2018**, *171*, 751–760. [CrossRef]

41. Sarmah, T.; Dhiman, S.K. Design and analysis of solar cabinet dryer for drying of potatoes. In *Advances in Engineering Design: Select Proceedings of FLAME 2018*; Prasad, A., Gupta, S.S., Tyagi, R.K., Eds.; Springer: Singapore, 2019; pp. 105–115. [CrossRef]

42. Kondareddy, R.; Sivakumaran, N.; Nayak, P.K. Drying kinetics of black turmeric (Curcuma caesia) with optimal controller assisted low cost solar dryer. *Food Res.* **2019**, *3*, 373–379. [CrossRef]

43. Shamekhi-Amiri, S.; Gorji, T.B.; Gorji-Bandpy, M.; Jahanshahi, M. Drying behaviour of lemon balm leaves in an indirect double-pass packed bed forced convection solar dryer system. *Case Stud. Therm. Eng.* **2018**, *12*, 677–686. [CrossRef]

44. Bhardwaj, A.K.; Kumar, R.; Chauhan, R. Experimental investigation of the performance of a novel solar dryer for drying medicinal plants in Western Himalayan region. *Sol. Energy* **2019**, *177*, 395–407. [CrossRef]

45. Poonia, S.; Singh, A.K.; Jain, D. Design development and performance evaluation of photovoltaic/thermal (PV/T) hybrid solar dryer for drying of ber (Zizyphus mauritiana) fruit. *Cogent Eng.* **2018**, *5*, 1507084. [CrossRef]

46. Poblete, R.; Cortes, E.; Macchiavello, J.; Bakit, J. Factors influencing solar drying performance of the red algae Gracilaria chilensis. *Renew. Energy* **2018**, *126*, 978–986. [CrossRef]

47. Sreekumar, A.; Rajarajeswari, K. Accelerated food processing through solar drying system. *IOP Conf. Ser. Mater. Sci. Eng.* **2018**, *377*, 012218. [CrossRef]

48. Macías-Ganchozo, E.R.; Bello-Moreira, I.P.; Trueba-Macías, S.L.; Anchundia-Muentes, X.E.; Anchundia-Muentes, M.E.; Bravo-Moreira, C.D. Design, development and performance of solar dryer for pineapple (*Ananas comosus* (L.) Merr.), mamey (*Mammea americana* L.) and banana (Musa paradisiaca L.) fruit drying. *Acta Agron.* **2018**, *67*, 30–38. [CrossRef]

49. Wanyama, J.; Galyaki, C.; Muyonga, J.; Kiggundu, N.; Banadda, N. Design and CFD simulation of a parabolic solar fruit dryer. In Proceedings of the 10th International Conference on Simulation and Modelling in the Food and Bio-Industry 2018, FOODSIM 2018, Ghent, Belgium, 8–12 April 2018; EUROSIS-ETI: Ostend, Belgium, 2018; pp. 301–303.

50. Tanongkankit, Y.; Narkprasom, K.; Narkprasom, N. Empirical modeling on hot air drying of fresh and pre-treated pineapples. In Proceedings of the 3rd International Conference on Chemical and Food Engineering, Tokyo, Japan, 8–9 April 2016; EDP Sciences: Les Ulis, France, 2016; p. 02007. [CrossRef]

51. Silva, K.S.; Garcia, C.C.; Amado, L.R.; Mauro, M.A. Effects of edible coatings on convective drying and characteristics of the dried pineapple. *Food Bioprocess Technol.* **2015**, *8*, 1465–1475. [CrossRef]

52. Talbot, P.; Lhote, M.; Heilporn, C.; Schubert, A.; Willaert, F.-X.; Haut, B. Ventilated tunnel solar dryers for small-scale farmers communities: Theoretical and practical aspects. *Dry. Technol.* **2016**, *34*, 1162–1174. [CrossRef]

53. Srisittipokakun, N.; Kirdsiri, K. Drying pineapple using a mix mode solar dryer. *Adv. Mater. Res.* **2014**, *979*, 11–15. [CrossRef]

54. Gudiño-Ayala, D.; Calderón-Topete, Á. Pineapple drying using a new solar hybrid dryer. *Energy Procedia* **2014**, *57*, 1642–1650. [CrossRef]

55. Moumegni-Kameni, C. Development of a Concept of Utilization of Renewable Energy Sources in Developing Countries. Bachelor's Thesis, Augsburg University of Applied Sciences, Augsburg, Germany, 2015. (In German)

56. Lamidi, R.O.; Jiang, L.; Wang, Y.D.; Pathare, P.B.; Roskilly, A.P. Techno-economic analysis of a biogas driven poly-generation system for postharvest loss reduction in a Sub-Saharan African rural community. *Energy Convers. Manag.* **2019**, *196*, 591–604. [CrossRef]

57. Drigalski, L. Development and Practical Application of a Model of Pineapple Solar Dryer Taking into Account the Most Important Parameters and Economic Factors. Master's Thesis, Augsburg University of Applied Sciences, Augsburg, Germany, 2016. (In German).

58. DIN EN 322. *Wood-Based Panels; Determination of Moisture Content*; Beuth Verlag: Berlin, Germany, 1993. (In German)

59. DIN EN 15414-3. *Solid Recovered Fuels—Determination of Moisture Content Using the Oven Dry Method—Part 3: Moisture in General Analysis Sample*; Beuth Verlag: Berlin, Germany, 2011. (In German)

60. Durakovic, B. Design of experiments application, concepts, examples: State of the art. *PEN Period. Eng. Nat. Sci.* **2017**, *5*, 421–439. [CrossRef]

61. Meteotest, A.G. *Meteonorm 7 User Guide*; Meteotest AG: Bern, Switzerland, 2016.

62. Kleppmann, W. *Design of Experiments. Optimization of Products and Processes*; Hanser: München, Germany, 2013. (In German)

63. Stieglitz, R.; Heinzel, V. *Thermal Solar Energy. Basics, Technology, Applications*; Springer: Berlin, Germany, 2012. (In German)

64. Hossain, M.D.; Bala, B.K.; Hossain, M.A.; Mondol, M.R.A. Sorption isotherms and heat of sorption of pineapple. *J. Food Eng.* **2001**, *48*, 103–107. [CrossRef]

65. Heimbrock, M. Optimization of an Air Collector for the Solar Drying of Tropical Fruits. Bachelor's Thesis, Augsburg University of Applied Sciences, Augsburg, Germany, 2016. (In German)

Article

Changing the Day-Ahead Gate Closure to Wind Power Integration: A Simulation-Based Study

Hugo Algarvio [1,2,*], António Couto [1], Fernando Lopes [1] and Ana Estanqueiro [1]

[1]   LNEG–National Laboratory of Energy and Geology, 1649-038 Lisbon, Portugal
[2]   Instituto Superior Técnico, 1049-001 Lisbon, Portugal
*   Correspondence: hugo.algarvio@tecnico.ulisboa.pt; Tel.: +351-210-924-600

Received: 5 June 2019; Accepted: 16 July 2019; Published: 18 July 2019

Abstract: Currently, in most European electricity markets, power bids are based on forecasts performed 12 to 36 hours ahead. Actual wind power forecast systems still lead to large errors, which may strongly impact electricity market outcomes. Accordingly, this article analyzes the impact of the wind power forecast uncertainty and the change of the day-ahead market gate closure on both the market-clearing prices and the outcomes of the balancing market. To this end, it presents a simulation-based study conducted with the help of an agent-based tool, called MATREM. The results support the following conclusion: a change in the gate closure to a time closer to real-time operation is beneficial to market participants and the energy system generally.

Keywords: day-ahead market; balancing market; gate closure; forecast uncertainty; wind power forecast; agent based simulation; the MATREM system

---

1. Introduction

Electricity market (EMs) are a complex and continuously evolving reality—new players are emerging and new strategic behaviors are gaining more active roles, meaning that researchers and practitioners did not yet solve the problems associated with this new reality [1,2]. Chief among these problems are the ones related to the increase in non-dispatchable renewable generation, or variable renewable energy (VRE), such as solar and wind power. VRE is characterized by substantial investment costs, but near-zero marginal costs, and great variability, thus increasing the uncertainty of the net load. VRE is normally the marginal resource, since it is operated at maximum capacity (taking into account the weather conditions). These characteristics have a strong influence on the outcomes of energy markets, reducing market-clearing prices [3]. Accordingly, existing market designs should be analyzed to determine if they are still efficient to deal with high levels of VRE (see, e.g., [4–6]).

The question of a suitable day ahead market design for a better integration of VRE has been the subject of a great deal of research (see, e.g., [7–10]). For the particular case of an effective gate closure, the potential solution discussed by the research community involves its adjustment to a time closer to the first hour of delivery. The main reason behind this adjustment is related to the forecast of renewable generation, which typically presents large errors for long time horizons, due to the stochastic nature of the atmosphere. Since the day-ahead market (DAM) closes typically at 12:00 p.m. (CET), the bids of wind power producers need to be performed through power forecasts computed at least 12 to 36 h ahead. Consequently, the adjustment of the gate closure is very important to enable a fair participation of VRE in EMs, since all energy producers (dispatchable and non-dispatachable) participate in the day-ahead market under the same rules. In case the energy production differs from the commitments resulting from the DAM, the differences should be balanced in the intraday market and/or the balancing market.

In 2014, the International Agency of Energy (IEA) presented a report about energy markets and renewable generation [11], scoring key EM features according to eight dimensions: non-VRE dispatch,

VRE dispatch, dispatch interval, last schedule update, grid representation, interconnector management and system services definition and market. Half of these dimensions are directly related to the improvement of forecasts, namely dispatch interval, grid representation, interconnector management and last schedule update. IEA considers that all physical transactions should be performed through implicit auctions in a centralized pool without feed-in-tariffs (FiTs) or other incentives to renewable energy sources (RES), with an interval up to 10 min using updates up to 30 min before real-time operation. In addition, the reserve requirements should be computed stochastically, taking into account different scenarios for the share of VRE, and be remunerated through marginal pricing. Furthermore, dispatch intervals should be reduced—that is, should be as short as possible—to enable VRE producers to perform high accurate forecasts. Such market modifications can contribute to a new paradigm—a paradigm of VRE integration without FiTs, where VRE investors do not have a guaranteed return, being remunerated by the market price, and subject to the payment of penalties for their deviations.

In 2016, the European Union (EU) presented the "Clean Energy for all Europeans", a package of measures to promote the integration of renewable generation and to harmonize the European markets. In 2017, the EU presented a new proposal for regulating the pan-European market [12]. Article 7 considers that market operators should trade as close as possible to real-time operation, and no later than the gate closure of the intraday cross-zonal market. In [13], we presented an overview of the potential effects of the new EU proposal on the integration of renewable generation. In [14–16], we analyzed the impact of both high levels of renewable generation and significant forecast errors on the outcomes of the DAM. As a preliminary result, we conclude that a gate-closure closer to real time operation is beneficial to wind power producers.

This article builds on our previous work on market design. It considers a specific market design element, namely the gate closure, and investigates how changes in this element can better accommodate the increasing levels of renewable generation. Specifically, this article analyzes the impact of both wind power forecast uncertainty and a change in the gate closure of the DAM, from 12:00 p.m. to 2:00 p.m. (CET), on the day-ahead market prices and also on the outcomes of the balancing market. The proposed gate closure encompasses two perspectives. The first is related to technical requirements. In a power system with a significant number of conventional power plants, there is a need of a certain lead-time to adjust generation levels in a cost efficient manner [17]. The second perspective is related to the most reliable meteorological information to feed the wind power forecast systems.

This article presents a simulation-based study conducted with the help of an agent-based tool. Multi-agent systems (MAS) are a somewhat new area of study (see, e.g., [2,18]). MAS are fundamentally coupled networks of computational agents that cooperate to resolve issues that are over their individual competence. Theoretically, MAS are an optimal fit for the distributed structure of liberalized energy markets. Accordingly, the work presented here makes use of a multi-agent simulator for competitive energy markets, called MATREM [19,20] (MATREM stands for Multi-agent Trading in Electricity Markets). The following aspects are examined in the paper, in order to assess the benefits of postponing the gate closure of the day-ahead market:

- The influence of the forecast accuracy on the day-ahead market, namely on the level of prices, and price volatility;
- The influence of the forecast accuracy on the balancing reserve requirements.

In addition, the following questions are addressed in the paper:

- How can forecast improvements possibly reduce the total balancing reserve requirements?
- Which parts of the system cause the need for a balancing demand? In addition, what is the associated energy quantity?

The remainder of the article is structured as follows. Section 2 discusses the role of wind power forecasts in energy markets. Section 3 presents an overview of the MATREM system, focusing on the day-ahead and the balancing markets. Section 4 describes the method considered in the experimental work, highlighting the main tasks and the relationships between them. Section 5 presents the case study and discusses the simulation results. Lastly, in Section 6, some conclusions are drawn.

2. The Role of Wind Power Forecast in Day-Ahead and Balancing Markets

In order to maintain high standards of service quality, in particular in what regards the security of supply and the system robustness, the system operator must be aware of the current and future values of wind power for each area and connection points of the grid [21]. Currently, an efficient and safe operation of power systems [21,22] requires that wind power production be well forecasted, and, when coupled with a load forecast system, both should enable the reduction of the need to balance the energy in the reserve markets, usually at high costs. During the past few years, numerous approaches have been developed for wind power forecast based on numerical weather prediction (NWP) models. Comprehensive reviews are presented in the literature (see, e.g., [23,24]). NWP models, as the Fifth-Generation Penn State/NCAR Mesoscale Model (MM5) [25], resolve the formulations that rule the state of the atmosphere using numerical methodologies.

Notwithstanding the advances on NWP, systematic errors still persist due to the inherent chaotic behaviour of the atmosphere, in which minor errors, at an early stage, will increase in the deterministic chaotic system. For a long time horizon, this situation could result in a large deviation of the forecast when compared with the observation [26,27]. This drawback can be partly explained by the physical formulation and parametrization of the atmosphere processes, the initial and boundary conditions (IC), among others. In fact, as indicated by several authors, one of the main sources of error and uncertainty, when numerical mesoscale models are applied, is derived from the ICs that feed the model, which are essentially atmospheric information provided by analysis/forecast products. Indeed, several authors have shown that these data have a crucial impact on the mesoscale model outcomes [28–30]. ICs are a three-dimensional set of meteorological data to force the boundary conditions of the model, and together with a terrain and roughness database, enable for conducting numerical physical simulations, for the region under analysis, in a time horizon comprising the day-ahead market. The ICs are obtained from global atmospheric models, such as the global forecast model system (GFS), with both a low time (6 h: at 12:00 a.m., 6:00 a.m., 12:00 p.m. and 6:00 p.m. UTC) and low spatial resolutions (e.g., 50 km) [31].

With the increasing levels of VRE generation in EMs, the underlying impact of wind power forecast into the power system has been explored by several authors. For instance, in [27], the authors studied the certainty gain effect of a wind power producer that participates in the day-ahead market, by delaying the forecasts according to NWP data availability. The results obtained demonstrate that NWP data availability determines the wind power forecast accuracy in the day-ahead market. In [6], by using a case study where wind parks are allowed to participate in the Portuguese tertiary reserve, the authors concluded that changing the market time unit from 1 h to 15 min reduces their imbalances about 10%. Thus, schedule updates as close as possible to real-time operation may strongly reduce the effect of forecast errors. However, the majority of the physical transactions of energy are performed in the day-ahead market (around 90% in Europe [32]), meaning that intraday markets (in Europe) and real-time markets (in the United States and Australia) have been less attractive, despite the fact that they allow schedule updates closer to real-time operation. Balancing markets operate essentially in real time and have attractive prices, when compared to spot markets, but have less liquidity (smaller trading quantities). A study conducted in Denmark [33] concluded that the participation of wind power producers in balancing markets increases the wind energy value [34] only 4.5%. However, some preliminary results (see, e.g., [35]) indicated that a suitable day-ahead market design can contribute to a large increase in the wind energy value, mainly by reducing the penalties paid with deviations.

This article explores the adjustment of the DAM gate-closure considering wind power forecasts according to the typical availability of the IC meteorological data. Currently, the Portuguese wind farm producers may participate in the Iberian day-ahead market by considering forecasts based on meteorological information obtained 18 h before the first trading hour (see Figure 1). To accomplish the technical requirements, such as the unit commitment and the time needed to perform all the steps necessary to obtain the wind power forecasts, the proposed new gate closure time is set to 2:00 p.m. (considering the 12:00 p.m. IC conditions). This adjustment can be favourable to deal with the variability of stochastic energy sources [36–38].

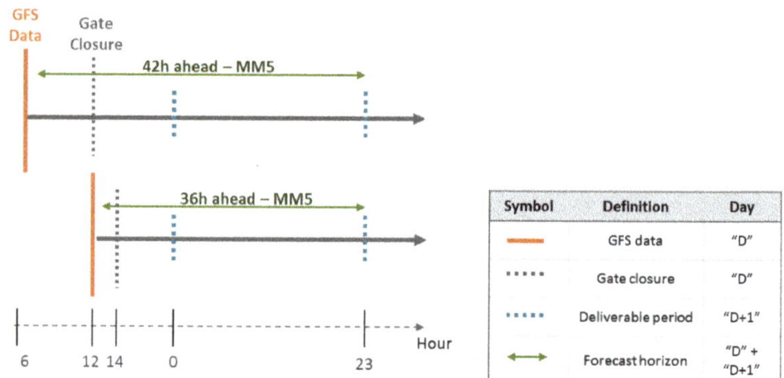

Figure 1. Global Forecast System (GFS) data availability and DAM design: (**top**)—actual; (**bottom**)—proposed in this work.

3. The MATREM Simulator

The major components of the MATREM system include a day-ahead market, an intra-day market, a futures market, a balancing market, and a marketplace for negotiating the terms and conditions of "tailored" bilateral contracts [19,20]. The system supports seven types of market entities: generating companies (GenCos), retailers (RetailCos), aggregators of VRE, coalitions of consumers, traditional consumers, market operators and system operators. All entities are modeled as software agents.

GenCo agents may own one power plant or a set of power plants with different technologies. Typically, they sell energy in the day-ahead market and the intra-day market—that is, the centralized markets—as well as in the bilateral market—the futures market. RetailCo agents buy energy from GenCos in the centralized markets as well as in the futures market, and subsequently, re-sell that energy to private consumers (by signing bilateral contracts with them). Aggregators of VRE allow the participation of wind power produces and other types of VRE producers in the centralized markets. Coalitions of consumers are essentially alliances of end-use consumers with the goal of reducing their energy cost, typically by increasing their bargaining power. Large traditional consumers can trade energy in the centralized markets and the bilateral market, while small traditional consumers may ally into coalitions or establish private bilateral contracts with retailers.

The day-ahead market is cleared one day in advance—that is, in day D for each of the 24 h of the next day (D + 1), as illustrated in Figure 1. The intraday market is a short-run market involving several auction sessions. Typically, both markets operate according to the system marginal pricing algorithm, although MATREM also supports locational marginal pricing. Supply-side agents compete by submitting offers to sell energy while demand-side agents submit offers to buy energy. The system ranks the selling offers by increasing price and the buying offers by decreasing price, obtaining the supply and demand curves, respectively. Next, the market operator computes the market-clearing prices and sends the results to the system operator, who checks (in a preliminary way) the security constraints.

The stability of the power system is a task associated with the system operator. To this end, this agent needs to take access to reserve capacity for the provision of system services. MATREM considers three different types of reserve capacity: primary reserve (or frequency control reserve), secondary reserve (or fast active disturbance reserve), and tertiary reserve (or slow active disturbance reserve). Tertiary reserve has to be available within 15 min and its activation is carried out manually—this type of reserve is the most important for the work described here.

Tertiary reserve is traded by the system operator in a day-ahead tender. This agent defines the needs for up and downregulation, receives the offers from the balance responsible parties, and computes the market-clearing prices by using a simplified version of the system marginal pricing algorithm. Two different simulations are performed, one for computing the upregulation price, and another for determining the downregulation price. Following the clearing process, the system operator can perform an imbalance settlement process.

The futures market is an organized market for trading standard bilateral contracts. Such contracts are agreements by which the parties take on the obligation to buy or to sell electricity, in a standardized quantity and quality, on a predefined date and place, at a price agreed in the present. The bilateral marketplace allows private parties to negotiate the terms and conditions of tailored (or customized) long-term contracts, specifically forward contracts [39] and contracts for difference [40]. To this end, market participants are equipped with a model that handles two-party and multi-issue negotiation [41,42].

4. Method

This work makes use of a diverse range of models and involves a set of tasks. Figure 2 depicts a flowchart of the approach considered, highlighting the main tasks and the relationships between them.

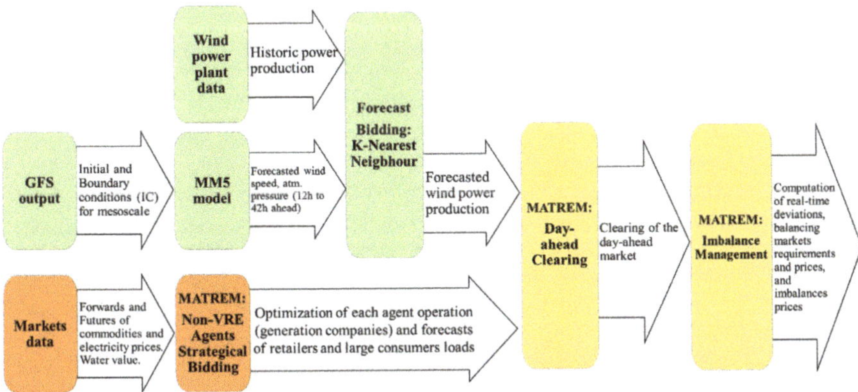

Figure 2. Overview of the methodology considered in this work, specifically the method and the main tasks related to the definition of the bids of the wind power producer (green) and the other market participants (light orange), as well as the procedure associated with the operation of the day-ahead market and the imbalance management process (light yellow).

4.1. Wind Power Forecast Approach

In the last few years, one of the key scientific research topics is the development of reliable forecasting systems. These systems are being widely employed in different fields, such as the prediction of consumer comfort based on temperature and humidity, electricity demand, wind speed and direction for onshore wind power, and also metocean conditions for offshore wind power, and irradiation for solar power (see, e.g., [43–46]).

For the particular case of wind power forecast systems, several techniques have been proposed in the literature (see, e.g., [47–52]). To support the participation of wind power producers in the DAM, existing wind power forecast systems are generally based on statistical post processing approaches coupled with mesoscale NWP outputs [23]. In this article, a K-nearest neighbour (K-NN) methodology is applied to provide the wind power forecasts. This statistical approach, also known as analogous forecast, has been applied by several authors due to its simplicity, effectiveness and non-parametric features [53,54].

The K-NN methodology is considered a lazy learning methodology, since there is no need to (previously) build a model or a function. Instead, the methodology uses only similar situations from the historical data to forecast. This characteristic is especially suitable to forecast weather dependent variables, such as wind power. The K-NN methodology considers that atmospheric flow releases the same local impact. Consequently, and taking into account the atmospheric flow characteristics, where the meteorological weather patterns have a tendency to repeat over certain regions, from time to time, the wind power forecast production for a determined hour can be determined from a similar meteorological weather pattern from historical events. The K-NN technique has shown a high efficiency in forecasting phenomena where the synoptic variability is predominant, such as precipitation [55] and wind speed [56]. Currently, this technique has already been explored by several authors for forecasting wind power, showing good performance [57–59]. To estimate the degree of similarity with each historical event, the Euclidean distance, with a trajectory matrix providing information regarding the state of the atmosphere on the preceding times, is the most common metric used in the wind forecast sector [59].

Due to a large number of degrees of freedom of the atmosphere, it is usual to apply a principal component analysis (PCA) approach to the input data. This procedure allows to filter atmospheric perturbation that represents only background noise [27]. To assist in this filtering process, the North criterion [60] was applied, allowing to identify the appropriate number of principal components (PCs) to be used in each meteorological parameter input.

Figure 3 depicts the flowchart of the K-NN methodology applied in this work. In order to calibrate the forecasting methodology, sensitivity studies were performed, using two years of real wind power data from a set of wind parks located in the central region of Portugal. These sensitive tests comprise the suitable number of K nearest neighbours, the size of the trajectory matrix and the selection of the most adequate meteorological parameters. After some tests, the most adequate configuration of the forecast methodology was to set K to 10 using a trajectory matrix with a lag of 3 h, composed by the first three PCs of the longitudinal component of the wind, wind speed, and atmospheric pressure. Thus, the deterministic wind power forecast is based on the average value of the ten historic events.

The normalized root mean square error (NRMSE) represents the quadratic difference between the estimated value (based on the proposed methodologies), and was normalized by the nominal capacity of wind park: P^{nom}. The correlation r is used in statistics to measure how strong the relationship is between two variables, in this case, the observed power P_t^{obs} and the forecast power P_t^{for}. The F Test is used to verify the statistical significance between the deviations associated with the two different gate-closures considered in this work. Mathematically, the previous parameters are defined by the following equations:

$$NRMSE\,[\%] = \sqrt{\frac{\sum \left(P_t^{obs} - P_t^{for} \right)^2}{N}} \times \frac{1}{P^{nom}} \tag{1}$$

$$r = \frac{cov\left(P_t^{obs}, P_t^{for} \right)}{\sqrt{var\left(P_t^{obs} \right) var\left(P_t^{for} \right)}} \tag{2}$$

$$F = \frac{var\left(P_t^{obs} - P_t^{for1} \right)^2}{var\left(P_t^{obs} - P_t^{for2} \right)^2} \tag{3}$$

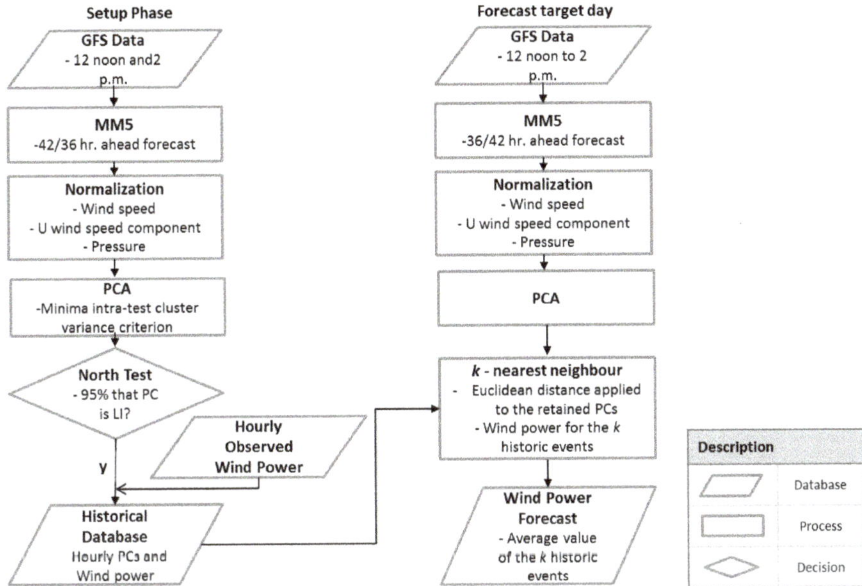

Figure 3. Wind power forecasts methodology based on NWP data coupled with a K-NN approach.

4.2. Selection of Representative Days

The underlying goal of choosing representative days is to statistically detect the most typical wind power daily patterns and, at the same time, clustering together days that exhibit identical patterns. This procedure allows for: (i) feeding the MATREM simulator without resorting to extensively time-consuming simulations, and (ii) assessing, for instance, the typical profiles that can jeopardize the wind power producer revenues enabling for adopting measures to mitigate their risk exposure. The identification of representative days is a suitable approach widely employed to increase the knowledge of a determined parameter allowing the creation of decision support systems (e.g., the classification of the type of profile of electricity customers [61]). A K-medoids clustering algorithm [62] is used in this work to find the representative days based on the daily observed wind power profile for the wind parks considered. This technique allows arranging the input data with similar characteristics into clusters in order to achieve the most representative daily profile. With this step, it is possible to identify (in a statistical way) statistically independent patterns from the data that can be related to physical processes. Clustering algorithms are unsupervised learning processes typically applied to find and split the data according to the similarity among the observations, in a way that is always closer to the elements of the same cluster, and dissimilar among the remaining clusters [62,63].

The main advantages of the K-medoids algorithm, when compared to others non-hierarchical clustering algorithms (e.g., the K-means algorithm) are: (1) to be more robust to noise and outliers by using the median values, and (2) to allow for selecting data points as centres (medoids) [63]. The K-medoids technique used in this work is classified as a non-hierarchical clustering algorithm, allowing to group the data into K clusters. The suitable K, i.e., the number of clusters, is predetermined through the Calinski–Harabasz (CH) criterion [64].

The wind power input matrix ($X_{d,t}$) for the clustering algorithm is defined as follows:

$$X_{d,t} = \begin{vmatrix} Z_{1,1} & Z_{1,2} & \cdots & Z_{1,t} \\ Z_{2,1} & Z_{2,2} & \cdots & Z_{2,t} \\ \vdots & \vdots & \vdots & \vdots \\ Z_{d,1} & Z_{d,2} & \cdots & Z_{d,h} \end{vmatrix} \tag{4}$$

where t represents the wind power observed during a predetermined hour of day d during the period 2009–2010.

4.3. Measures of Economic Results

In this section, the formulation used to compare the results between the 12:00 p.m. scenario (hereafter designated as the *base case*) and 2:00 p.m. scenario (hereafter designated as the *upgraded case*) is presented. The total remuneration (per hour) of the wind power producers is as follows:

$$R_t = \sum_{t=0}^{23} P_t^{bid} C_t^{dayahead} + \begin{cases} \left(P_t^{obs} - P_t^{bid} \right) C_t^{updeviation}, & \text{for } P_t^{bid} < P_t^{obs} \\ \left(P_t^{obs} - P_t^{bid} \right) C_t^{downdeviation}, & \text{for } P_t^{bid} > P_t^{obs} \end{cases} \tag{5}$$

where:

- $C_t^{dayahead}$ is the day-ahead price at hour t;
- $C_t^{updeviation}$ and $C_t^{downdeviation}$ are the up and down deviation costs, respectively.

It is important to note that $P_t^{bid} C_t^{dayahead}$ is the part consisting of the remuneration obtained from the day-ahead. The other part consists of the remuneration obtained from the deviations. Therefore, the total remuneration R_d and average remuneration $\bar{R}_d$ for a specific day d are as follows:

$$R_d = \sum_{t=0}^{23} R_t \tag{6}$$

$$\bar{R}_d = \frac{R_d}{\sum_{t=0}^{23} P_t} \tag{7}$$

The average remuneration obtained in the day-ahead market $\bar{R}_d^{dayahead}$ is defined as follows:

$$\bar{R}_d^{dayahead} = \frac{\sum_{t=0}^{23} P_t^{bid} C_t^{dayahead}}{\sum_{t=0}^{23} P_t^{bid}} \tag{8}$$

In addition, the average remuneration obtained by considering the deviations $\bar{R}_d^{deviation}$ is as follows:

$$\bar{R}_d^{deviation} = \begin{cases} \dfrac{\sum_{t=0}^{23} \left(P_t^{obs} - P_t^{bid} \right) C_t^{updeviation}}{\sum_{t=0}^{23} \left(P_t^{obs} - P_t^{bid} \right)}, & \text{for } P_t^{bid} < P_t^{obs} \\[4mm] \dfrac{\sum_{t=0}^{23} \left(P_t^{obs} - P_t^{bid} \right) C_t^{downdeviation}}{\sum_{t=0}^{23} \left(P_t^{obs} - P_t^{bid} \right)}, & \text{for } P_t^{bid} > P_t^{obs} \end{cases} \tag{9}$$

With these formulae, it is possible to compute the average wind power value, the energy transacted in the tertiary reserve market, and the tertiary reserve cost for both scenarios. Both the reserve cost and the electric system levelized cost are computed by taking into account the occurrence of each representative day and the traded energy. In order to assess the gain effect regarding the proposed adaptation of the gate closure of the day-ahead market, several key performance indicators (KPIs) are defined (see Table 1).

Table 1. KPIs considered in this work.

KPI	Objective	Formulation
Increase in the wind power value to the market	Increase the remuneration of wind power producers in the market	Equation (7)
Increase in the forecast accuracy	Reduce the forecast error	Equation (3)
Reduction of total control reserve by wind power forecast improvements	Reduce the balance needs of the tertiary reserve	Equations (1) and (2)
Reduction of the tertiary reserve costs	Reduce the balance costs of the tertiary reserve	Tertiary reserve simulation
Reduction of total operating costs in the electricity system by forecast improvements	Reduce the system costs with the day-ahead market and the tertiary reserve	Day-ahead market and tertiary reserve simulations

5. The Case Study

This section describes a case study to analyse the effect of wind power forecasts errors on the outcomes of the DAM. The following two scenarios are considered: (i) a base scenario, where the DAM closes at 12:00 p.m. (the bids of the wind power producers are based on a wind forecast performed 18 to 42 h ahead), and (ii) an updated scenario, where the DAM closes at 2:00 p.m. (the bids of WPPs are based on an updated forecast performed 12 to 36 h ahead).

5.1. Software Agents and Wind Power Profiles

This study makes use of data published by the Iberian electricity market (MIBEL) and involves the simulation of the day-ahead market prices as well as the balancing market prices. Market participants are modeled as software agents, defined with the help of the MATREM system. Since the normal operation of the daily market of MIBEL involves a number of bids on the order of thousands for a particular hour, there is a need to make some simplifications related to the number of software agents, in order to avoid a large computational complexity. Accordingly, the main agents considered in the study are as follows: a market operator (S1), a system operator (S2), twelve producers (supply-side agents) and four retailers (demand-side agents). Table 2 presents the characteristics of the supply-side agents. The wind aggregator (agent P_1) represents the Portuguese wind farms.

Table 2. Producer agents (software agents) and their key characteristics.

Agent	Country	Technology	Maximum Capacity (MW)	Marginal Cost (€/MWh)
P1	Portugal	Wind	2500	0
P2	Portugal	Renewable mix	2000	0
P3	Portugal	Hydroelectricity	4500	[30; 60]
P4	Portugal	Coal	1800	≈30
P5	Portugal	Combined Cycle Gas	3000	≈55
P6	Portugal	Fuel oil	2000	≈70
P7	Spain	Renewable mix	30,000	0
P8	Spain	Hydroelectricity	16,500	[30; 60]
P9	Spain	Coal	10,000	≈30
P10	Spain	Nuclear	7500	≈30
P11	Spain	Combined Cycle Gas	22,000	≈55
P12	Spain	Fuel oil	4000	≈70

It is important to note that the detection of violations related to the interconnection constraints between Portugal and Spain leads to a process of market splitting, resulting in different price areas for Portugal and Spain. After a careful examination, we concluded that this situation applies for some hours of the days under consideration. In practice, this means two sets of simulation for each hour of operation, one for Portugal and another for Spain. However, for convenience, and in the interests of simplicity, the day-ahead market is cleared for Portugal only.

The forecast methodology is deterministic and uses the following: (i) numerical weather prediction data outputs (see Section 4), and (ii) observed data for a set of wind farms during the period 2009–2010. The wind farms have a nominal capacity of 250 MW (10% of the Portuguese installed capacity, in 2010). This value is upscaled to 2500 MW to obtain a meaningful impact on the market results. The observed wind power profiles are depicted in Figure 4a. The representability of each wind power profile is also shown in Figure 4b.

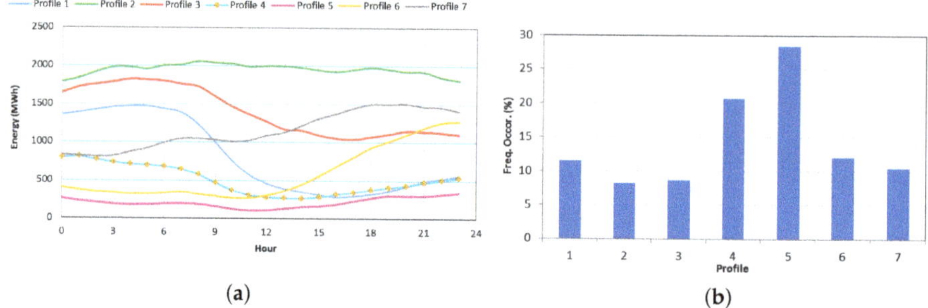

Figure 4. Wind energy typical profile (**a**) and representability of each wind power profile during the 2-year period of the study (**b**).

The analysis of Figure 4 supports the consideration that the Portuguese wind farms are located in a mountain region, since several wind power profiles show the typical features of wind speed in such a region (although with different intensity)—that is, due to the thermal stratification and local effects [65], the highest wind speed is associated with the nocturnal period. Moreover, the most common wind power profile shows a reduced production during all day. On the other hand, profile 2 is associated with a high level of wind power production, which occurs during the passage of severe meteorological phenomena, as the cyclone systems [66]. This profile shows the lowest number of occurrences.

5.2. Wind Power Forecast Deviations

Figure 5 depicts the wind power forecast deviations (forecast minus observed production) for the seven representative days, at 12:00 p.m. (base case, left) and 2:00 p.m. (updated case, right). The figure shows that the wind power forecast deviations at 12:00 p.m. have an absolute value that is almost twice that of the deviations at 2:00 p.m. Moreover, wind power fluctuations are considerably higher in the base scenario. For instance, the uncertainty in profile 2 ranges between −200 and 1700 MW.

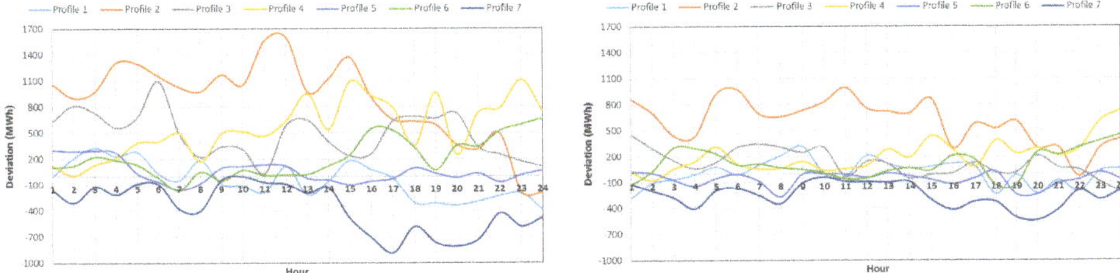

Figure 5. Wind power forecast deviations at 12:00 p.m. (base case, **left**) and at 2:00 p.m. (updated case, **right**).

Table 3 depicts the forecast results regarding both the NRMSE and the correlation values between the two scenarios. The most significant improvement was observed in profile 2. This profile, usually associated to extreme weather conditions, shows a strong improvement in the correlation and the NRMSE values. The link between wind power variability as well as the uncertainty with extreme weather conditions was described by several authors [23,66–69], who demonstrated that larger errors in the wind power forecast are expected to occur during severe weather conditions with strong dynamics (e.g., storms and cold fronts) when compared with weather conditions associated with stationary systems (e.g., anticyclonic systems). For a 0.05 significance level, the critical value is 2.01, which means that comparing both deviations (see Figure 5), exist statistically differences in all profiles, except in profile 6 (see Table 3).

Profile 4 also shows a strong improvement in the wind power forecast for the upgrade scenario. Profile 5 shows the lowest NRMSE amelioration that can be explained with the capabilities to obtain a reliable forecast during calm wind speed conditions [67,69]. Consequently, results from Table 3 highlighted the fact that the data considered about the ICs strictly define the wind power forecast errors. Thus, as expected, since the most up-to-date information on the state of the atmosphere is used, postponing the gate closure can strongly reduce the uncertainty of the bids that the wind power producers submit to the day-ahead market.

Table 3. Correlation and NRMSE between the observed and forecasted wind power and F value of the wind power deviations for each scenario and wind power profile.

Parameter Difference	Simulation	Wind Power Profile						
		1	2	3	4	5	6	7
Correlation	Base Scenario	0.91	−0.35	0.72	0.54	0.79	0.78	0.47
	Upgrade Scenario	0.95	0.27	0.89	0.86	0.92	0.88	0.71
NRMSE (%)	Base Scenario	4.23	19.46	10.72	12.45	2.73	5.99	8.94
	Upgrade Scenario	2.87	12.85	3.93	5.52	2.10	3.90	5.61
F	Both Scenario	2.07	3.39	2.61	2.81	2.37	1.94	3.85

5.3. Simulation Results

5.3.1. Impact of Wind Power Forecast on Market Outcomes

In a preliminary attempt to understand the relation of the aforementioned wind power forecast uncertainty results with the market outcomes, we observed that the deviations are essentially positive in most of the representative days (day seven is the exception), meaning that the forecasts normally underestimate the wind power values. From the point-of-view of wind power producers, this situation (underestimation) can be profitable, since wind power is offered at a price around 0 € in the day-ahead market, and thus an underestimation forecast will increase the market price. On the other hand, an overestimation forecast can decrease the day-ahead market price, decreasing the wind energy value. Therefore, it can be concluded that an underestimation of the wind power production in the day-ahead market (shortage forecast) overestimates the importance of the wind power and also gives an extra amount to the supply-side agents (by increasing the market-clearing price). On the other hand, an overestimation of the wind power production (excess forecast) undervalues the wind power value. Moreover, due to the payment of penalties in the balancing market, the latter situation can lead to a drastic reduction in the revenues of wind power producers. Thus, an underestimation of wind power production can be more profitable for wind power producers.

The results of the simulations for both the base scenario (12:00 p.m.) and the updated scenario (2:00 p.m.) are presented in Tables 4 and 5, respectively. In both tables, details regarding the outcomes of postponing the market closing time by two hours are also provided. As mentioned previously, the sum of the wind power deviations at 12:00 p.m. is almost twice the deviations associated with 2:00 p.m., and on some days it is more than twice the amount, such as the 3rd and 4th representative days. The results also show that the underestimation of wind power production can increase the market price and, as discussed previously, can overvalue the wind power. This conclusion takes into account the fact that the average day-ahead remuneration is higher at 12:00 p.m., in almost all days, excluding the 7th representative day. In this representative day, the overestimation of wind power production leads to an undervaluation of the wind power value. This behavior can also be observed in the revenue deviations results. In fact, when both day-ahead and revenue deviations are positive, an underestimation of wind power is obtained. Consequently, power producers receive a positive revenue for the extra energy at the real-time operation (normally less than in the day-ahead market, due to penalties). When the revenue deviations are negative, this normally means an overestimation of wind power production when compared with the forecast, so there will be a need to pay a value higher than the day-ahead price.

Table 4. Key results for the base scenario (12:00 p.m. scenario).

Profile	Power Forecast 12 h (Base Scenario)						
	1	2	3	4	5	6	7
Day-ahead energy bid (GWh)	22.01	25.93	22.18	14.37	10.58	8.06	12.61
Day-ahead Remuneration (k€)	767.21	951.6	885.58	529.22	389.25	317.03	525.92
Average Day-ahead Rem. (€/MWH)	34.85	36.7	39.93	36.82	36.81	39.31	41.72
Real time deviations (MWh)	4.45	21.86	11.61	13.26	2.55	5.5	8.78
Deviations Revenue (k€)	−143.6	460.33	226.29	317.29	9.94	98.66	−527.7
Average Remuneration (€/MWh)	29.94	30.03	32.91	30.63	33.75	31.75	−0.46
Total Remuneration (k€)	623.62	1411.9	111.19	846.51	399.19	415.69	−1.77

One of the key parameters that can be used to compare the base scenario with the updated scenario is the average remuneration of the wind power producers. Considering both tables, it can be seen that, in the updated scenario, the wind value (average remuneration of the wind power) is always higher, and the proposed market design element seems even more relevant when a wind power forecast overestimation occurs. For instance, for the 7th representative day, the average remuneration of the wind power producer is negative, in the base case. Therefore, if hypothetically wind power producers are active players in the market, with the current market design, when the day-ahead market prices are lower than the deviation prices, the producers who have an energy shortage (when compared to their energy forecast bids) may have a negative average remuneration.

Table 5. Key results for the updated scenario (2:00 p.m. scenario).

Profile	Power Forecast 2:00 p.m. (Updated Scenario)						
	1	2	3	4	5	6	7
Day-ahead energy bid (GWh)	20.62	32.48	30.83	22.59	13.41	10.26	9.83
Day-ahead Remuneration (k€)	699.61	1173.3	1187.7	784.42	488.76	400.31	414.5
Average Day-ahead Rem. (€/MWh)	33.93	36.12	38.53	34.73	36.44	39.03	42.17
Real-time deviations (MWh)	2.76	14.59	3.78	5.2	1.75	3.82	6.01
Deviations Revenue (k€)	−45.3	330.82	48.19	124.78	−81.56	43.28	−355.6
Average Remuneration (€/MWh)	31.42	31.99	36.58	32.9	34.43	33.88	15.41
Total Remuneration (k€)	654.31	1504.1	123.59	909.2	407.2	443.6	58.94

Now, taking into account the representability of each wind power profile during the two years of data, it is possible to compute the average wind energy value, the energy transacted in the tertiary reserve market, and the tertiary reserve costs, among other key parameters, for both the 12:00 p.m. and the 2:00 p.m. scenarios (see Table 6). The results in the Table show that the upgraded case leads to better results in almost all key indicators (the exception is the reserve cost).

Also, the results suggest that a reduction in the day-ahead market prices, due to a reduction in the forecast errors (NRMSE), leads to an increase in the wind power producers revenues, by allowing for reducing their losses associated with the deviation penalties. With a reduction of forecast errors, the quantity of reserve required to compensate the deviations decreases. However, both the reserve levelized cost and the reserve cost increased. These results are associated with a decrease in the system requirements for down reserve (see the reserve direction parameter). In this way, the system operator receives an inferior remuneration from the down reserve, which leads to an increase of the down reserve price. This increase, together with a decrease in the down reserve utilization and the day-ahead market prices, will negatively affect the revenue of the power plants that bid at the tertiary reserve market. This behavior is associated with a reduction of the remuneration of wind power producers from the day-ahead market since they need to pay a high price for the down reserve, for a small quantity of energy.

Table 6. Average values of the key parameters under evaluation per day.

Key Parameters	Simulations	
	12:00 p.m. **(Base Scenario)**	**2:00 p.m.** **(Updated Scenario)**
Wind Energy Value (€/MWh)	28.46	31.69
Forecast NRMSE (%)	8.03	4.52
Reserve Use (GWh)	11.1	6.23
Reserve Direction (GWh)	−7.44	−3.83
Reserve Costs (M€)	−0.1	−0.04
Reserve Levelized Cost (€/MWh)	−7.26	−3.83
Down Reserve Price (€/MWh)	20.07	25.7
Day-ahead Market Cost (M€)	18.75	18.63
Day-ahead Market Prices (€/MWh)	38.6	37.97
Day-ahead Market Energy (GWh)	487.71	491.31
Electric System Cost (M€)	18.65	18.58
Electric System Lev. Cost (€/MWh)	38.84	38.12

5.3.2. Quantifying the Gain Effect of the Proposed Market Design Change

The KPIs defined in Section 4 enable for quantifying the gain effect by setting the gate closure of the day-ahead market to 2:00 p.m., instead of 12:00 p.m. (see Table 7). The change of the day-ahead market closing time brings benefits to the system in general, with a reduction around 16.5% in the total costs. As stated before, the wind power producers and the demand-side players benefit from this change. Wind power producers gain from selling the same quantity at a higher net price (market price with fewer penalties). The demand-side players gain from buying a similar quantity of electricity at a lower price. The system operator benefits from using less the reserve market to balance the system (a reduction around 44%) and also receives the lowest revenue (less 56%) from the down reserve market (part of the tertiary reserve market), which means that the agents that deviate will pay higher penalties. Notwithstanding, the power producers that buy energy from the tertiary reserve market (down reserve) decrease their revenues, by having less energy to buy at a higher price.

A comparison of the results shown in Table 7 with the main results presented by the literature (see Section 2) allows us to conclude that postponing the gate-closure of the day-ahead market only two hours seems to be a change of market design that can bring large benefits to power systems generally.

Table 7. Key performance indicators (KPIs).

KPIs (%)	Changing from 12:00 p.m. to 2:00 p.m.
Increase in the wind power value to the market	11.34
Increase in the forecast accuracy	43.71
Reduction of total control reserve by forecast improvements in wind power	43.87
Reduction of the tertiary reserve costs	−56.25
Reduction of total operating costs in the electricity system by forecast improvements	16.46

6. Conclusions

The article analyzed a change in the gate closure of the day-ahead market to deal with the uncertainty of variable generation. Specifically, it considered the adjustment of the day-ahead market closing time from 12:00 p.m. (currently in use in most of the European electricity markets) to 2:00 p.m. (CET). To test this adjustment, a case study based on real data from a set of aggregated wind parks in Portugal, and also data from the supply-side (producers) and demand-side (retailers buying electricity for the end-use consumers) of the Iberian Market (MIBEL), as an approximation of the entire system, was established.

Wind power forecast data were obtained using a K-NN approach based on data from a NWP model. The day-ahead market was simulated using the system marginal pricing algorithm for seven representative days, taking into account two scenarios: gate closure of the day-ahead market set to 12:00 p.m. (base case) and to 2:00 p.m. (updated case). The seven representative wind power production days enable: (i) to simulate only the most common typical wind power production days in the region under study, and (ii) to identify the wind power profiles that can jeopardise the revenues of the wind power producers or can pose serious challenges to transmission system operators.

From a wind power forecast perspective, the results show that some wind power profiles clearly benefit from a change in the market design. Regarding the electricity market perspective, the results show that the change of the day-ahead market closing time to 2:00 p.m. benefits the wind power producers at both a technical and financial level by decreasing the forecast errors and increasing the revenues. The consumers can also potentially take advantage of this change due to a potential reduction in the overall system costs, which may allow a reduction of the electricity tariffs. The system operator benefits from a reduction in the wind park forecast errors, by reducing the requirements to maintain the production/demand balance (technical benefit), requiring less energy from the tertiary reserve market. However, they also receive less money from the downregulation (financial loss), which means that the agents that deviate will need to pay higher penalties to compensate this loss. The power producers that bid at the down reserve market have a loss due to a decrease in the system requirements for this type of reserve, i.e., the system operator requires less down reserve quantity to balance the system, which increases the price of this market.

The results presented in this work highlight that electricity markets with high shares of VRE could benefit from DAMs with a gate closure closer to real-time operation, due to improvements in the forecast accuracy. The full integration of wind power in markets can be possible with substantial changes to the current market designs, especially in power systems with a high share of VRE integration, as expected in the forthcoming years with the society decarbonization.

Author Contributions: H.A. and A.C. conceived and designed the experiments and wrote a preliminary version of the article. In particular, A.C. conceived and designed the forecast models and computed the forecasts. H.A. designed and conducted the simulations with the help of the MATREM system. F.L. performed a deep revision of the earlier versions of the article, regarding both language and scientific content. In addition, F.L. developed the MATREM system (thus enabling performing all the simulations). A.E. coordinated the whole process.

Funding: This work was supported by FCT (Fundação para a Ciência e Tecnologia) under grant agreement PD/BD/105863/2014 (H. Algarvio).

Conflicts of Interest: The authors declare no conflict of interest.

Abbreviations

CH	Calinski–Harabasz
EM	Electricity market
EU	The European Union
GenCos	Generating companies
GFS	Global Forecast System
IC	Initial and boundary conditions
K-NN	K-nearest neighbour
KPI	Key performance indicator
MAS	Multi-agent system
MATREM	The Multi-agent Trading in Electricity Markets
MIBEL	The Iberian electricity market
MM5	The Fifth-Generation Penn State/NCAR Mesoscale Model
MO	Market operator
NRMSE	Normalized root mean square error
NWP	Numerical weather prediction
PC	Principal component
PCA	Principal component analysis
RetailCos	Retailers
SMP	System marginal pricing
TSO	Transmission system operator
VRE	Variable renewable energy

Indices

d	Day
t	Time period (hour)

Parameters

P^{nom}	Nominal capacity

Variables

$C_t^{dayahead}$	Day-ahead price
$C_t^{downdeviation}$	Down deviation cost
$C_t^{updeviation}$	Up deviation cost
P_t^{bid}	Bidding power
P_t^{down}	Down reserve power
$P_t^{down,ref}$	Down reserve power for the reference case
P_t^{for}	The wind power forecast
P_t^{obs}	The hourly observed wind power production
P_t^{ref}	Reference bidding power
P_t^{up}	Up reserve power
$P_t^{up,ref}$	Reference up reserve power
R	Remuneration
$X_{d,t}$	Wind power input matrix

References

1. Conejo, A.J.; Carrión, M.; Morales, J.M. *Decision Making under Uncertainty in Electricity Markets*; Springer: New York, NY, USA, 2010.
2. Lopes, F.; Coelho, H. *Electricity Markets with Increasing Levels of Renewable Generation: Structure, Operation, Agent-Based Simulation and Emerging Designs*; Springer International Publishing: Cham, Switzerland, 2018. doi:10.1007/978-3-319-74263-2
3. Ela, E.; Milligan, M.; Bloom, A.; Cochran, J.; Botterud, A.; Townsend, A.; Levin, T. Overview of Wholesale Electricity Markets. In *Electricity Markets with Increasing Levels of Renewable Generation: Structure, Operation, Agent-Based Simulation and Emerging Designs*; Springer International Publishing: Cham, Switzerland, 2018; pp. 3–21. doi:10.1007/978-3-319-74263-2_1

4. Moiseeva, E.; Wogrin, S.; Hesamzadeh, M.R. Generation flexibility in ramp rates: Strategic behavior and lessons for electricity market design. *Eur. J. Oper. Res.* **2017**, *261*, 755–771. [CrossRef]

5. Ela, E.; Milligan, M.; Bloom, A.; Botterud, A.; Townsend, A.; Levin, T. Incentivizing Flexibility in System Operations. In *Electricity Markets with Increasing Levels of Renewable Generation: Structure, Operation, Agent-Based Simulation, and Emerging Designs*; Springer International Publishing: Cham, Switzerland, 2018; pp. 95–127. doi:10.1007/978-3-319-74263-2_5

6. Algarvio, H.; Lopes, F.; Couto, A.; Estanqueiro, A. Participation of Wind Power Producers in Day-ahead and Balancing Markets: An Overview and a Simulation-based Study. *WIREs Energy Environ.* **2019**, *2019*, e343. [CrossRef]

7. Vilim, M.; Botterud, A. Wind power bidding in electricity markets with high wind penetration. *Appl. Energy* **2014**, *118*, 141–155. [CrossRef]

8. IRENA. *Adapting Market Design to High Shares of Variable Renewable Energy*; International Renewable Energy Agency: Abu Dhabi, UAE, 2017.

9. Roques, F.; Finon, D. Adapting electricity markets to decarbonisation and security of supply objectives: Toward a hybrid regime? *Energy Policy* **2017**, *105*, 584–596. [CrossRef]

10. Zhang, C.; Yan, W. Spot Market Mechanism Design for the Electricity Market in China Considering the Impact of a Contract Market. *Energies* **2019**, *12*, 1064. [CrossRef]

11. International Energy Agency. *The Power of Transformation: Wind, Sun and the Economics of Flexible Power Systems*; International Energy Agency: Paris, France, 2014.

12. European Commission. Proposal for a Regulation of the European Parliament and of the Council on the Internal Market for Electricity. February 2017. Available online: http://ec.europa.eu/energy/sites/ener/files/documents/1_en_act_part1_v9.pdf (accessed on 5 July 2019).

13. Algarvio, H.; Lopes, F.; Couto, A.; Santana, J.; Estanqueiro, A. Effects of Regulating the European Internal Market on the integration of Variable Renewable Energy. *WIREs Energy Environ.* **2019**, *2019*, e346. [CrossRef]

14. Algarvio, H.; Couto, A.; Lopes, F.; Estanqueiro, A.; Santana, J. Multi-Agent Energy Markets with High Levels of Renewable Generation: A Case-Study on Forecast Uncertainty and Market Closing Time. In Proceedings of the 13th International Conference Distributed Computing and Artificial Intelligence, Sevilla, Spain, 1–3 June 2016; pp. 339–347.

15. Algarvio, H.; Couto, A.; Lopes, F.; Estanqueiro, A.; Holttinen, H.; Santana, J. Agent-Based Simulation of Day-Ahead Energy Markets: Impact of Forecast Uncertainty and Market Closing Time on Energy Prices. In Proceedings of the 27th Workshop on Database and Expert Systems Applications (DEXA 2016), Porto, Portugal, 5–8 September 2016; pp. 166–170.

16. Algarvio, H.; Couto, A.; Lopes, F.; Estanqueiro, A.; Santana, J. Multi-agent Wholesale Electricity Markets with High Penetrations of Variable Generation: A Case-study on Multivariate Forecast Bidding Strategies. In *Highlights of Practical Applications of Cyber-Physical Multi-Agent Systems*; Springer: Cham, Switzerland, 2017; pp. 340–349.

17. Burgholzer, B.; Fontaine, A.; Galmiche, F.; Völler, S.; Jaehnert, S.; Chicharro, F.; Camacho, L.; Ahcin, P. D5.2 Report on the quantitative evaluation of policies for post 2020 RES-E targets. Market4RES Report. September 2016. Available online: http://market4res.eu/wp-content/uploads/D5-2_final-003.pdf (accessed on 17 July 2019).

18. Wooldridge, M. *An Introduction to Multiagent Systems*; Wiley: Chichester, UK, 2009.

19. Lopes, F. MATREM: An Agent-based Simulation Tool for Electricity Markets. In *Electricity Markets with Increasing Levels of Renewable Generation: Structure, Operation, Agent-Based Simulation and Emerging Designs*; Springer: Cham, Switzerland, 2018; pp. 189–225. doi:10.1007/978-3-319-74263-2_8

20. Lopes, F.; Coelho, H. Electricity Markets and Intelligent Agents. Part II: Agent Architectures and Capabilities. In *Electricity Markets with Increasing Levels of Renewable Generation: Structure, Operation, Agent-Based Simulation and Emerging Designs*; Springer, Cham, Switzerland, 2018; pp. 49–77. doi:10.1007/978-3-319-74263-2_3 [CrossRef]

21. Kahn, P. Numerical Techniques for Analysing Market Power in Electricity. *Electr. J.* **1998**, *11*, 34–43. [CrossRef]

22. Hobbs, F. Linear Complementarity Models of Nash-Cournot Competition in Bilateral and POOLCO Power Markets. *IEEE Trans. Power Syst.* **2001**, *16*, 194–202. [CrossRef]

23. Giebel, G.; Brownsword, R.; Kariniotakis, G.; Denhard, M.; Draxl, C. The State-of-the-Art in Short-Term Prediction of Wind Power: A Literature Overview. Deliverable Report D-1.2, Project ANEMOS.Plus, Contract 038692, DTU. 2011; p. 106. Available online: https://orbit.dtu.dk/fedora/objects/orbit:8339 7/datastreams/file_5277161/content (accessed on 17 July 2019).
24. Jung, J.; Broadwater, R. Current Status and Future Advances for Wind Speed and Power Forecasting. *Renew. Sustain. Energy Rev.* **2014**, *31*, 762–777. [CrossRef]
25. Grell, G.; Dudhia, J.; Stauffer, D.R. A Description of the Fifth-Generation Penn State/NCAR Mesoscale Model (MM5), NCAR Technical Note NCAR/TN-398+STR. 1995; p. 121. Available online: http://opensky.ucar.edu /islandora/object/technotes%3A170/datastream/PDF/view (accessed on 17 July 2019).
26. Lorenz, E.N. Deterministic Nonperiodic Flow. *J. Atmos. Sci.* **1963**, *20*, 130–141. [CrossRef]
27. Couto, A.; Rodrigues, L.C.; Costa, P.; Silva, J.; Estanqueiro, A. Wind power participation in electricity markets—The role of wind power forecasts. In Proceedings of the 16th IEEE International Conference on Environment and Electrical Engineering, Florence, Italy, 7–10 June 2016; p. 6.
28. Alvarez, I.; Gomez-Gesteira, M.; Carvalho, D. Deep-Sea Research II Comparison of different wind products and buoy wind data with seasonality and interannual climate variability in the southern Bay of Biscay (2000–2009). *Deep Sea Res. Part II* **2014**, *106*, 38–48. [CrossRef]
29. Wang, A.; Zeng, X. Evaluation of multireanalysis products with in situ observations over the Tibetan Plateau. *JGR Atmos.* **2012**, *117*, 1–12. [CrossRef]
30. Soukissian, T.H.; Papadopoulos, A. Effects of different wind data sources in offshore wind power assessment. *Renew. Energy* **2015**, *77*, 101–114. [CrossRef]
31. Rutledge, G.K.; Alpert, J.; Ebisuzaki, W. NOMADS: A Climate and Weather Model Archive at the National Oceanic and Atmospheric Administration. *Bull. Am. Meteorol. Soc.* **2006**, *87*, 327–341. [CrossRef]
32. ACER. Annual Report on the Results of Monitoring the Internal Electricity and Gas Markets in 2016. Electricity Wholesale Markets Volume. October 2017. Available online: http://www.acer.europa.eu/Official_document s/Acts_of_the_Agency/Publication/ACER%20Market%20Monitoring%20Report%202016%20-%20ELECTR ICITY.pdf (accessed on 5 July 2019).
33. Skytte, K.; Bobo, L. Increasing the value of wind: From passive to active actors in multiple power markets. *WIREs Energy Environ.* **2019**, *8*, doi:10.1002/wene.328. [CrossRef]
34. Mills, A.; Wiser, R. Changes in the economic value of wind energy and flexible resources at increasing penetration levels in the Rocky Mountain Power Area. *Wind Energy* **2014**, *17*, 1711–1726. [CrossRef]
35. Holttinen, H.; Miettinen, J.; Couto, A.; Algarvio, H.; Rodrigues, L. Wind power producers in shorter gate closure markets and balancing markets. In Proceedings of the 13th International Conference on the European Energy Market, Porto, Portugal, 6–9 June 2016; p. 5.
36. Holttinen, H.; Milligan, M.; Ela, M.; Menemenlis, N.; Dobschinski, J.; Rawn, B.; Bessa, R.J.; Flynn, D.; Gomez Lazaro, E.; Detlefsen, N. Methodologies to determine operating reserves due to increased wind power. *IEEE Trans. Sustain. Energy* **2012**, *3*, 713–723. [CrossRef]
37. Milligan, M.; Kirby, B.; Holttinen, H.; Kiviluoma, J.; Estanqueiro, A.; Martín-Martínez, S.; van Hulle, F.; Gomez-Lázaro, E.; Pineda, I.; Smith, J.C. *Wind Integration Cost and Cost-Causation*; National Renewable Energy Lab. (NREL): Golden, CO, USA, 2013; p. 6.
38. González-Aparicio, I.; Zucker, A. Impact of wind power uncertainty forecasting on the market integration of wind energy in Spain. *Appl. Energy* **2015**, *159*, 334–349. [CrossRef]
39. Algarvio, H.; Lopes, F.; Santana, J. Bilateral Contracting in Multi-agent Energy Markets: Forward Contracts and Risk Management. In *Highlights of Practical Applications of Agents, Multi-Agent Systems, and Sustainability: The PAAMS Collection (PAAMS 2015)*; Springer International Publishing: Cham, Switzerland, 2015; pp. 260–269.
40. Sousa, F.; Lopes, F.; Santana, J. Contracts for Difference and Risk Management in Multi-agent Energy Markets. In *Advances in Practical Applications of Agents, Multi-Agent Systems, and Sustainability: The PAAMS Collection (PAAMS 2015)*; Springer International Publishing: Cham, Switzerland, 2015; pp. 339–347.
41. Lopes, F.; Mamede, N.; Novais, A.Q.; Coelho, H. A Negotiation Model for Autonomous Computational Agents: Formal Description and Empirical Evaluation. *J. Intell. Fuzzy Syst.* **2002**, *12*, 195–212.
42. Lopes, F.; Coelho, H. Concession Behaviour in Automated Negotiation. In *E-Commerce and Web Technologies*; Springer: Berlin/Heidelberg, Germany, 2010; pp. 184–194. doi:10.1007/978-3-642-15208-5_17
43. Dong, Y.; Zhang, Z.; Hong, W.C. A hybrid seasonal mechanism with a chaotic cuckoo search algorithm with a support vector regression model for electric load forecasting. *Energies* **2018**, *11*, 1009. [CrossRef]

44. Capizzi, G.; Lo Sciuto, G.; Napoli, C.; Tramontana, E. Advanced and adaptive dispatch for smart grids by means of predictive models. *IEEE Trans. Smart Grid* **2017**, *9*, 6684–6691. [CrossRef]

45. Brusca, S.; Capizzi, G.; Lo Sciuto, G.; Susi, G. A new design methodology to predict wind farm energy production by means of a spiking neural network-based system. *Int. J. Numer. Model.* **2019**, *32*, e2267. [CrossRef]

46. Hong, W.C.; Li, M.W.; Geng, J.; Zhang, Y. Novel chaotic bat algorithm for forecasting complex motion of floating platforms. *Appl. Math. Model.* **2019**, *72*, 425–443. [CrossRef]

47. Lenzi, A.; Steinsland, I.; Pinson, P. Benefits of spatiotemporal modeling for short-term wind power forecasting at both individual and aggregated levels. *Environmentrics* **2018**, *29*, e2493. [CrossRef]

48. Ohba, M.; Kadokura, S.; Nohara, D. Medium-Range Probabilistic Forecasts of Wind Power Generation and Ramps in Japan Based on a Hybrid Ensemble. *Atmosphere* **2018**, *9*, 423. [CrossRef]

49. Agarwal, P.; Shukla, P.; Sahay, K.B. A Review on Different Methods of Wind Power Forecasting. In Proceedings of the 2018 International Electrical Engineering Congress (iEECON), Krabi, Thailand, 7–9 March 2018; pp. 1–4.

50. Liu, H.; Chen, C.; Lv, X.; Wu, X.; Liu, M. Deterministic wind energy forecasting: A review of intelligent predictors and auxiliary methods. *Energy Convers. Manag.* **2019**, *195*, 328–345. [CrossRef]

51. Zhang, J.; Cui, M.; Hodge, B.M.; Florita, A.; Freedman, J. Ramp forecasting performance from improved short-term wind power forecasting over multiple spatial and temporal scales. *Energy* **2017**, *122*, 528–541. [CrossRef]

52. Ellis, N.; Davy, R.; Troccoli, A. Predicting wind power variability events using different statistical methods driven by regional atmospheric model output. *Wind Energy* **2015**, *18*, 1611–1628. [CrossRef]

53. Martínez, F.; Pérez, M.D.; Frías, M.P.; Rivera, A.J. A methodology for applying k -nearest neighbor to time series forecasting. *Artif. Intell. Rev.* **2017**. [CrossRef]

54. Taneja, S.; Gupta, C.; Goyal, K.; Gureja, D. An Enhanced K-Nearest Neighbor Algorithm Using Information Gain and Clustering. In Proceedings of the 2014 Fourth International Conference on Advanced Computing & Communication Technologies, Rohtak, India, 8–9 February 2014; pp. 325–329.

55. Zorita, E.; Von Storch, H. The Analog Method as a Simple Statistical Downscaling Technique: Comparison with More Complicated Methods. *J. Clim.* **1999**, *12*, 2474–2489. [CrossRef]

56. Pascual, A.; Valero, F.; Martín, M.L.; Morata, A.; Luna, M.Y. Probabilistic and deterministic results of the ANPAF analog model for Spanish wind field estimations. *Atmos. Res.* **2012**, *108*, 39–56. [CrossRef]

57. Alessandrini, S.; Sperati, S.; Pinson, P. A comparison between the ECMWF and COSMO Ensemble Prediction Systems applied to short-term wind power forecasting on real data. *Appl. Energy* **2013**, *107*, 271–280. [CrossRef]

58. Martín, M.L.; Valero, F.; Pascual, A.; Sanz, J.; Frias, L. Analysis of wind power productions by means of an analog model. *Atmos. Res.* **2014**, *143*, 238–249. [CrossRef]

59. Junk, C.; Delle Monache, L.; Alessandrini, S.; Cervone, G.; von Bremen, L. Predictor-weighting strategies for probabilistic wind power forecasting with an analog ensemble. *Meteorol. Zeitschrift* **2015**, *24*, 361–379. [CrossRef]

60. North, G.; Bell, T. Sampling errors in the estimation of empirical orthogonal functions. *Mon. Weather Rev.* **1982**, *110*, 699–706. [CrossRef]

61. Ramos, S.; Duarte, J.; Soares, J.; Vale, Z.; Duarte, F. Typical Load Profiles in the Smart Grid Context—A Clustering Methods Comparison. In Proceedings of the IEEE Power and Energy Society General Meeting 2012, San Diego, CA, USA, 22–26 July 2012.

62. Park, H.S.; Jun, C.H. A simple and fast algorithm for K-medoids clustering. *Expert Syst. Appl.* **2009**, *36*, 3336–3341. [CrossRef]

63. Huth, R.; Beck, C.; Philipp, A.; Demuzere, M.; Ustrnul, Z.; Cahynová, M.; Kyselý, J.; Tveito, O.E. Classifications of atmospheric circulation patterns: Recent advances and applications. *Ann. N. Y. Acad. Sci.* **2008**, *1146*, 105–152. [CrossRef]

64. Calinski T.; Harabasz, J. A dendrite method for cluster analysis. *Commun. Stat.* **1974**, *3*, 27.

65. Draxl, C. On the Predictability of Hub Height Winds. Ph.D. Thesis, DTU Wind Energy, Roskilde, Denmark, 2012; p. 105.

66. Couto, A.; Costa, P.; Rodrigues, L.; Lopes, V.; Estanqueiro, A. Impact of Weather Regimes on the Wind Power Ramp Forecast in Portugal. *IEEE Trans. Sustain. Energy* **2015**, *6*, 934–942. [CrossRef]

67. Lange, M.; Heinemann, D. Relating the uncertainty of short-term wind speed predictions to meteorological situations with methods from synoptic climatology. In Proceedings of the European Wind Energy Conference EWEC, Madrid, Spain, 16–19 June 2003; p. 7.
68. Ernst, B.; Oakleaf, B.; Ahlstrom, M.L.; Lange, M.; Moehrlen, C.; Lange, B.; Focken, U.; Rohrig, K. Predicting the wind. *IEEE Power Energy Mag.* **2007**, *5*, 78–89. [CrossRef]
69. Trancoso, R. Operational Modelling as a Tool in Wind Power Forecast and Meteorological Warnings. Ph.D. Thesis, Instituto Superior Técnico, Lisbon, Portugal, 2012; p. 146.

MDPI

St. Alban-Anlage 66

4052 Basel

Switzerland

Tel. +41 61 683 77 34

Fax +41 61 302 89 18

www.mdpi.com

Energies Editorial Office

E-mail: energies@mdpi.com

www.mdpi.com/journal/energies